CAMBRIDGE COUNTY GEOGRAPHIES

General Editor: F. H. H. GUILLEMARD, M.A., M.D.

CUMBERLAND

Cambridge County Geographies

CUMBERLAND

by

J. E. MARR, Sc.D. F.R.S.

With Maps, Diagrams and Illustrations

Cambridge :
at the University Press
1910

CAMBRIDGE UNIVERSITY PRESS
Cambridge, New York, Melbourne, Madrid, Cape Town,
Singapore, São Paulo, Delhi, Mexico City

Cambridge University Press
The Edinburgh Building, Cambridge CB2 8RU, UK

Published in the United States of America by Cambridge University Press, New York

www.cambridge.org
Information on this title: www.cambridge.org/9781107660199

First published 1910
First paperback edition 2013

A catalogue record for this publication is available from the British Library

ISBN 978-1-107-66019-9 Paperback

PREFACE

I DESIRE to acknowledge my indebtedness to several friends and to certain publications.

I have naturally consulted the various writings of the late Chancellor R. S. Ferguson, and have got much information from the *Transactions* of the Cumberland and Westmorland Antiquarian and Archaeological Society. I have also obtained information about the "Worthies of Cumberland" from Mr H. Lonsdale's book bearing that title.

I wish to thank Mr T. Gray, Librarian of Tullie House, Carlisle, for much help when using the Bibliotheca Jacksoniana in that Institution. To Mr K. J. J. Mackenzie, M.A., I am indebted for information concerning the agriculture of the county. The Headmaster of Blencowe Grammar School, Mr F. B. Sandford, has kindly read the proofs.

I am, as ever, grateful for the courtesy of all connected with the University Press with whom I have been brought in contact while the work was in progress, and must specially thank Mr H. A. Parsons who undertook the production of the diagrams at the end of the book.

J. E. M.

31 *August*, 1910.

CONTENTS

ILLUSTRATIONS

MAPS

The illustrations on pp. 4, 20, 30, 40, 43, 48, 49, 90, 102, 160 are from photographs by Messrs Abraham and Sons of Keswick; those on pp. 6, 22, 39, 54, 82, 98, 99 by Mr Brunskill of Bowness-on-Windermere; those on pp. 8, 10, 12, 26, 28, 44, 46, 47, 50, 52, 57, 63, 105, 109, 127, 131, 143, 146, 149, 152, 165 by Mr Herbert Bell of Ambleside, that on p. 143 being a reproduction from the portrait by R. B. Haydon; those on pp. 24, 100, 112, 115, 116, 118, 125, 126, 144, 163 by Messrs F. Frith & Co. of Reigate; Mr F. W. Tassell of Carlisle supplied the photographs reproduced on pp. 92, 121; and Messrs W. H. Moss and Sons, Ltd., Whitehaven, those on pp. 34, 35, 86, 159; the portraits of Southey and Dalton are from photographs by Mr Emery Walker.

The illustration on p. 129 is reproduced from a figure in the *Proceedings* of the Cumberland and Westmorland Antiquarian and Archaeological Society by kind permission of the Officers; the illustrations on pp. 104, 106, 107, 119, 123, 156, 162 all from the Cumberland volume of Lysons' *Magna Britannia* (1816) and that on p. 111 from T. Rose's *Cumberland, Westmorland*, etc. The illustration on p. 98 from a photograph of an object in the Kendal Museum is reproduced by permission of the Museum Council.

1. County and Shire. Cumberland: Origin of the Word.

All Englishmen are proud of their country, and know some of the reasons which led to the growth of the English nation and caused its people to occupy that particular tract of country which they actually inhabit. Each of us, further, is proud of his native county. Many people of all ranks for example, young and old, take an interest in the annual struggle of the counties for supremacy in cricket. Yet comparatively few know the events which have caused our country to be separated into those divisions which we term counties. The irregular boundaries of these counties, which are so great a stumbling-block to the young student of geography, suggest that the causes which lead to the making of a county are by no means simple. At the present day, when divisions of a tract of land are made, they are often very simple. Look at the line which divides Canada from the United States. For a long stretch it is straight. Most of the states and counties in America are bounded by straight lines. So in our country new towns like Barrow-in-Furness and Middlesborough are built with most of the streets in straight lines running at right angles to each other. In these cases the whole scheme of the parcelling out of the tracts is thought out before the division is made. But in the case of our counties there was no such principle of arrangement. They gradually grew up under varying conditions, and the boundaries were shifted more than once. These boundaries have usually been determined by some physical feature of the country which could be readily utilised, and often formed an actual barrier between adjacent divisions. As we shall see later, the county of Cumberland is separated from the adjoining counties along most parts of its borders by hill-ridges or by streams. Many divisions of the tract of country which we call Cumberland were made before its present boundaries were fixed.

All of us must have remarked that the names of many counties end in "shire," as Lancashire, while others, as Cumberland and Westmorland, Kent and Essex, have not this ending. Shires are tracts of land which were divided by the Anglo-Saxons, the name itself being Anglo-Saxon, and meaning that the tract is due to the "shearing" or cutting up of larger tracts. Cumberland was never a shire, for it did not become part of England until William Rufus took possession of it in 1092, and it was not until early in the twelfth century that Henry I divided the old Cumbrian kingdom into two counties, Carleolum and Westmarieland, not differing very widely from the present Cumberland and Westmorland. The term county is from the old French word *comté*, a province governed by a count (*comes*), and it did not come into use till after the Norman Conquest. Such counties as Essex, Kent, and Sussex have kept their names, and roughly their boundaries as well, from the earliest times, and are survivals of former kingdoms.

The situation of Cumberland, which was for long on the borderland between Scotland and England, was the cause of frequent changes in the boundaries of the political divisions of the area. The more important changes will be noted when we consider the history of the area, but in the meantime we must remember that the county in its present condition only came into being at a late date as compared with many other counties.

Long after the Anglo-Saxon invasion the ancient British people maintained their independence in the western parts of the island in three kingdoms, West Wales, North Wales, and Strathclyde; the first named was the territory now occupied by Devon, Cornwall, and part of Somerset, the second by the present Wales, and the third by a part including Lancashire, Westmorland, Cumberland, and part of Scotland. The name Welshmen simply implies foreigners, and was used by the invaders to designate those of a different race from themselves; the British people called themselves Cymry. This name, the meaning of which is not known, was latinised into Cumbri by the Anglo-Saxon chronicler Ethelwerd (*died* 998?) and at the end of the ninth century the name Cumbria was applied to the territory of which the present Cumberland is a part. When Henry I made Cumberland and Westmorland into two counties, the former was at first called Carliol or Carleolum, after its principal city, but in 1177 it became the county of *Cumberland*, and has since retained that title.

2. General Characteristics. Position and Natural Conditions.

Cumberland, far as it is from the great centres of population has, nevertheless played considerable part in the political and military affairs of the past, chiefly because of

Angle Tarn and Bowfell

the situation of its most important town, Carlisle, close to the Scottish border. Most of it is, as it has been in the past, essentially pastoral, but of recent years the discovery of important deposits of coal and iron has made portions of the county important from the industrial point of view. Not very densely populated, on account of the hilly nature of a considerable portion of the county, the people in the past have lived mainly in scattered home-steads or small hamlets in the valley bottoms, or in larger villages and small towns of the lowland tracts. Away from the industrial centres there are no large towns.

Physically, the county may be divided into three main regions which we shall describe more fully in a subsequent chapter. The southern boundary from near Millom to the foot of Ullswater forms the south side of a roughly semicircular fell tract with radiating valleys, in some of which large lakes are situated. North of this is a horse-shoe-shaped tract of low ground, which bordering the coast from Millom to the Solway, sweeps inland up the Eden valley to Penrith and sends off a spur north-east-ward to the Cheviots. Lastly, to the east of this tract is another fell district, whose east side is the eastern part of the county boundary, while the west side runs nearly parallel. This tract is chiefly a portion of the Pennine range, but north of the river Irthing forms the western portion of the Cheviots.

Cumberland is a maritime county, inasmuch as its south-western boundary and a considerable part of its north-western limit, consist of coast-line with several harbours. Most of the rivers are small, swift and clear, and apart from estuarine portions are all unnavigable, save the Eden, which allows of the passage of small craft to a point a little below Carlisle.

There are no forests as the word is now understood; there are forests, it is true, such as Skiddaw Forest, but they are treeless. A forest is literally an open hunting

Wasdale Church and Great Gable

ground, and that was the nature of the forests of Cumberland. Woods do occur, but they are not numerous, and most of them have been planted in recent times. But there is still much coppice in the lower parts of the valleys, with thick growths of hazel, birch, willow, alder, ash, and oak. In former times there were true wooded forests, which have been destroyed. The most important was Inglewood Forest, occupying an area of over 150 square miles and stretching S., S.W., and S.E. for a distance of about 16 miles with a breadth of about 10 miles to Penrith. This was one of the royal forests in the fourteenth century.

The climate is mild, and the rainfall is rather high, though, as we shall see in a later chapter, there has been much exaggeration concerning the amount of rain which falls in the Lake District.

The scenery of the county is varied, and much of it is very beautiful. The fell region of the south-western portion is especially fine, and the great scarp of the Pennine chain overlooking the lowlands of the Eden valley is impressive. From a picturesque point of view the palm must be given to the valleys which, with their admixture of crag, bracken-clad slope and coppice, are very lovely. The waterfalls are miniature, but picturesque. The lakes present different types of beauty according to their surroundings. Less familiar, but worthy of particular notice, is the scenery of the estuaries of the Duddon and Esk. It is essentially of Lakeland type, and the sands add to the beauty of the scene. There is much variety also in the river-valleys. The open Vale of Eden, the

Lodore

burns of the Cheviots, the becks of the Pennines, and the various dales of the Lakeland portion, all have their characteristic features.

Apart then from what Cumberland has had to do with the development of the kingdom, it is a county which well deserves our regard from its physical characteristics.

3. Size. Shape. Boundaries.

Cumberland with Westmorland and the Furness district of Lancashire form an upland region between the Pennine hills on the east and the Irish Sea on the west, with the Solway and the river Liddell on the north, and Morecambe Bay on the south.

The greatest length of the county is 75 miles measured from Scotch Knowe to Hodbarrow Point; and the smallest breadth along a line taken through the heart of the county is 25 miles between the Solway and Penrith. The county encloses an area of 973,086 acres or 1520 square miles.

There are only eight English counties which are larger than Cumberland. It occupies about one thirty-third of the entire area of England.

Comparing it with the English counties which border it, we find that Northumberland and Lancashire are somewhat larger, Durham and Westmorland considerably smaller, and Yorkshire nearly four times as large.

The shape of Cumberland is very irregular, but if we leave the indentations out of account it forms a wide

The Foot of Loweswater

parallelogram with the angles at Kershope Burn, the river Tees near Cross Fell, Haverigg Point near Millom, and St Bees Head.

The eastern boundary between the first two angles named does not run north and south, but from 30° W. of N. to 30° E. of S.; the next side is almost from N.E. to S.W.; the third is nearly parallel to the first ; and the fourth nearly parallel to the second.

It will be advisable to follow the county limits carefully upon the map, and the variations in the heights should be noted, for the nature of this boundary is of great importance as bearing upon the history of Cumberland.

Along the eastern boundary Cumberland is in contact with Northumberland and Durham.   From its northern end at Scotch Knowe on Kershope Burn, the boundary runs south-eastward over the Cheviots to a tributary of the Irthing, descending tributary and main stream to near Gilsland station, where it ascends another tributary and thence takes a curve over the Pennine fells to include the Alston Moor district until it reaches the Tees at the junction of Cumberland, Westmorland, and Durham.

Along the second line it ascends to Tees Head, crosses the Pennines to Crowdundale Beck and descends this to its junction with the Eden, which it descends until the Eamont enters from the west.   The boundary then runs westward up this river to the foot of Ullswater, and takes the middle line of that lake to some distance below the lake-head, where it passes up Glencoin Beck to the summit ridge of the Helvellyn range, north of Helvellyn top.   It then follows the watershed first south and then south-west to Wrynose Pass, where Cumberland, West-morland, and Lancashire meet. It descends the Duddon from the Pass to the sea near Millom.

The third side of the parallelogram between Millom and St Bees is bordered by the sea.

The River Duddon

(Boundary between Cumberland and Lancashire)

The fourth line also has the sea (with the Solway estuary) as its boundary to the mouth of the essentially Scotch river Sark. It ascends this river for a few miles, then crosses eastward by an ancient earthwork (the Scots dike) to the Liddell, which it ascends to the junction of Kershope Water, up which it is continued to Scotch Knowe, the point from which we started.

4. Geology and Soil.

Before giving further account of the physical geography of the county it is necessary to learn somewhat of its geology, as the physical conditions are to a large extent dependent upon geological structure.

By Geology we mean the study of the rocks, and we must at the outset explain that the term *rock* is used by the geologist without any reference to the hardness or compactness of the material to which the name is applied; thus he speaks of loose sand as a rock equally with a hard substance like granite.

Rocks are of two kinds, (1) those laid down mostly under water, (2) those due to the action of heat.

The first kind may be compared to sheets of paper one over the other. These sheets are called *beds*, and such beds are usually formed of sand (often containing pebbles), mud or clay, and limestone, or mixtures of these materials. They are laid down as flat or nearly flat sheets, but may afterwards be tilted as the result of movement of the earth's crust, just as you may tilt sheets of paper, folding them into arches and troughs, by pressing them at either end. Again, we may find the tops of the folds so produced worn away as the result of the wearing action of rivers, glaciers, and sea-waves upon them, as one might cut off the tops of the folds of the paper with a pair of shears. This has happened with the ancient beds forming parts of the earth's crust, and we therefore often find them tilted, with the upper parts removed. Tilted beds are said to *dip*, the direction of dip being that in which the beds plunge *downwards*, thus the beds of an arch dip *away from* its crest, those of a trough *towards* its middle. The dip is at a low angle when the beds are nearly horizontal, and at a high angle when they approach the vertical position. The horizontal line at right angles to the direction of the dip is called the line of *strike*. Beds form strips at the surface, and the portion where they appear at the surface is called the *outcrop*. On a large scale the direction of outcrop generally corresponds with that of the strike. Beds may also be displaced along great cracks, so that one set of beds abuts against a different set at the sides of the crack, when the beds are said to be *faulted*.

The other kinds of rocks are known as igneous rocks, which have been melted under the action of heat and become solid on cooling. When in the molten state they have been poured out at the surface as the lava of volcanoes, or have been forced into other rocks and cooled in the cracks and other places of weakness. Much material is also thrown out of volcanoes as volcanic ash and dust, and is piled up on the sides of the volcano. Such ashy material may be arranged in beds, so that it partakes to some extent of the qualities of the two great rock groups.

The production of beds is of great importance to geologists, for by means of these beds we can classify the rocks according to age. If we take two sheets of paper, and lay one on the top of the other on a table, the upper one has been laid down after the other. Similarly with two beds, the upper is also the newer, and the newer will remain on the top after earth-movements, save in very exceptional cases which need not be regarded by us here, and for general purposes we may regard any bed or set of beds resting on any other in our own country as being the newer bed or set.

The movements which affect beds may occur at different times. One set of beds may be laid down flat, then thrown into folds by movement, the tops of the beds worn off, and another set of beds laid down upon the worn surface of the older beds, the edges of which will abut against the oldest of the new set of flatly deposited beds, which latter may in turn undergo disturbance and removal of their upper portions.

Again, after the formation of the beds many changes may occur in them. They may become hardened, pebble-beds being changed into conglomerates, sands into sandstones, muds and clays into mudstones and shales, soft deposits of lime into limestone, and loose volcanic ashes into exceedingly hard rocks. They may also become cracked, and the cracks are often very regular, running in two directions at right angles one to the other. Such cracks are known as *joints*, and the joints are very important in affecting the physical geography of a district. As the result of great pressure applied sideways, the rocks may be so changed that they can be split into thin slabs, which usually, though not necessarily, split along planes standing at high angles to the horizontal. Rocks affected in this way are known as *slates*.

If we could flatten out all the beds of England, and

arrange them one over the other and bore a shaft through them, we should see them on the sides of the shaft, the newest appearing at the top and the oldest at the bottom. Such a shaft would have a depth of between 50,000 and 100,000 feet. The beds are divided into three great groups called Primary or Palaeozoic, Secondary or Mesozoic, and Tertiary or Cainozoic, and at the base of the Primary rocks are the oldest rocks of Britain, which form as it were the foundation stones on which the other rocks rest, and are termed Precambrian rocks. The three great groups are divided into minor divisions known as systems.

In the following table (p. 17) a representation of the various great subdivisions or 'systems' of the beds which are found in the British Islands is shown. The names of the great divisions are given on the left-hand side, in the centre the chief divisions of the rocks of each system are enumerated, and on the right-hand the general characters of the rocks of each system are given.

With these preliminary remarks we may now proceed to a brief account of the geology of the county, though to render it intelligible we must also say something about the geology of the adjoining tracts of Westmorland and of the Furness district of Lancashire.

In the county of Cumberland the following systems are found and are represented on the geological map at the end of the book: Recent and Pleistocene, Jurassic, Triassic, Permian, Carboniferous, Silurian and Ordovician. The figure (p. 18) shows what is called a geological section, drawn across the county from near Silecroft to the county boundary north of Longtown and gives the

Names of Systems		Subdivisions	Characters of Rocks
TERTIARY	Recent Pleistocene	Metal Age Deposits Neolithic ,, Palaeolithic ,, Glacial ,,	Superficial Deposits
	Pliocene	Cromer Series Weybourne Crag Chillesford and Norwich Crags Red and Walton Crags Coralline Crag	Sands chiefly
	Miocene	Absent from Britain	
	Eocene	Fluviomarine Beds of Hampshire Bagshot Beds London Clay Oldhaven Beds, Woolwich and Reading Thanet Sands [Groups	Clays and Sands chiefly
SECONDARY	Cretaceous	Chalk Upper Greensand and Gault Lower Greensand Weald Clay Hastings Sands	Chalk at top Sandstones, Mud and Clays below
	Jurassic	Purbeck Beds Portland Beds Kimmeridge Clay Corallian Beds Oxford Clay and Kellaways Rock Cornbrash Forest Marble Great Oolite with Stonesfield Slate Inferior Oolite Lias—Upper, Middle, and Lower	Shales, Sandstones and Oolitic Limestones
	Triassic	Rhaetic Keuper Marls Keuper Sandstone Upper Bunter Sandstone Bunter Pebble Beds Lower Bunter Sandstone	Red Sandstones and Marls, Gypsum and Salt
PRIMARY	Permian	Magnesian Limestone and Sandstone Marl Slate Lower Permian Sandstone	Red Sandstones and Magnesian Limestone
	Carboniferous	Coal Measures Millstone Grit Mountain Limestone Basal Carboniferous Rocks	Sandstones, Shales and Coals at top Sandstones in middle Limestone and Shales below
	Devonian	Upper Mid } Devonian and Old Red Sand- Lower stone	Red Sandstones, Shales, Slates and Lime- stones
	Silurian	Ludlow Beds Wenlock Beds Llandovery Beds	Sandstones, Shales and Thin Limestones
	Ordovician	Caradoc Beds Llandeilo Beds Arenig Beds	Shales, Slates, Sandstones and Thin Limestones
	Cambrian	Tremadoc Slates Lingula Flags Menevian Beds Harlech Grits and Llanberis Slates	Slates and Sandstones
	Pre-Cambrian	No definite classification yet made	Sandstones, Slates and Volcanic Rocks

arrangement of the rocks in the county. It represents what would be seen on the sides of a deep cutting made through the county. In this section the smaller subdivisions of the systems which are found in the county are also indicated.

The actual Lake District consists chiefly of a mass of slaty rocks which occupy a tract of country of a roughly circular form about thirty miles in diameter. These rocks consist of three main groups, each composed of beds

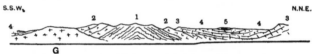

Geological Section from near Silecroft to the County Boundary N. of Longtown

(*Length about 70 miles*)

5	Lias		
4	Trias		
3	Carboniferous	G	Granite
2	Volcanic Series	} Ordovician	
1	Skiddaw Slates		

many thousands of feet in thickness. The oldest group, called the Skiddaw slates, is formed chiefly of clay slates with a few sandstone bands. The middle group, the Borrowdale Volcanic Series, is made up of lavas and ashes of various characters, while the upper group is very variable. Like the lower it consists of sediments, with impure limestones at the bottom followed by clay shales and sandstones. The rocks are very old and are known as the Ordovician and Silurian rocks, which are amongst the oldest of the British Isles or indeed of the world.

The masses of igneous rock which have been injected into the Ordovician rocks between Buttermere and Black Combe, and in a smaller degree around the Vale of St John, are probably of late Ordovician age. They are of a granitic character. The upper and lower groups contain many fossils. These fossils are interesting for several reasons, among others, because they give us some idea of the forms of life which existed in these early times—forms differing in many ways from those of the present day. Interesting as these fossils are, however, they are of slight importance as regards the geography of Cumberland, and we may pass them by with the bare notice of their occurrence.

These old slates of the Lake District after their formation were thrown into a great arch with the centre of the arch passing through the Skiddaw group of hills and the southern part of the arch sloping southward, so that the beds sink down into the ground as a whole in a southerly direction. Accordingly the oldest group of slates is found chiefly in the northern part of the district, the middle group in the central tract, and the youngest group in the southern portion.

At the time of the formation of the great arch the topmost rocks of the arch were removed by rivers, and probably by the waves of the sea, and a comparatively flat tract was formed by these processes of planing down. At the same time the rocks were much compressed and hardened, and the finer ones turned into slates. Some igneous rocks were at this date squeezed into them and cooled.

Derwentwater, Skiddaw, and Bassenthwaite from Falcon Crag

(Showing conical hills formed of Skiddaw slates)

On the levelled surface of these old rocks another set of beds was formed, belonging to what is known as the Carboniferous system, so called because it contains workable coal.

The most interesting of the lower beds are the masses of white limestone, which from its frequent existence in hills is spoken of as the Mountain Limestone. This contains a great number of fossils in places, and is indeed largely composed of them.

After the formation of the Carboniferous rocks, another movement took place, which we need not notice fully, as it was less important than the earlier one. At this period was probably injected a mass of igneous rock known as the "Whin Sill," lying nearly level among the limestones of the Pennine Hills.

Another set of rocks consisting chiefly of red sandstone was laid upon the older rocks. These belong to the New Red Sandstone age, so called to distinguish it from that of the Old Red Sandstone, which is practically unrepresented in Cumberland.

Long after this the Lake District was affected by another uplift, which was not saddle-shaped but dome-shaped, and as the upper rocks were again swept away we find the old slate rocks now on the surface in the centre of the dome with the Mountain Limestone forming an almost complete ring around them, and outside this ring of limestone is another, less perfect, of New Red Sandstone. On the east side of the New Red Sandstone of the Eden valley the Carboniferous rocks again appear at the surface in the Pennine hills, for the Eden valley

Scawfell

(Showing Crags formed of Volcanic Rocks)

between the Lake District and the Pennine chain is a geological trough.

With these descriptions and study of the geological map we can understand the structure of Cumberland which contains the north-western part of the dome, the northern part of the Eden valley trough, and part of the Pennine uplift.

The south-western part of the county is chiefly formed of the old slate rocks, which may be traced along the county boundary with Westmorland, from the estuary of the Duddon in the south-west to the foot of Ullswater on the north-east. Here occurs the junction with the ring of Carboniferous beds which form the strip of country on the west side of the Vale of Eden to Hesket Newmarket, where they turn westward and continue along the tract north of the slaty rocks of the Lake District, bearing south-westward near Cockermouth and disappearing where covered by New Red Sandstone rocks near Egremont. They reappear from under these rocks near Haverigg Point and at Millom form the extreme south of the county, but disappear below the waters of the Duddon estuary, to come out on its eastern side in Lancashire.

The New Red Sandstone occurs in Cumberland in the lower part of the Eden valley and in the valley of the Petteril as a comparatively narrow strip between Penrith and Wreay station. At Wreay it expands and forms a triangular area whose eastern side is bounded by the Pennine and Cheviot Hills, the southern side by the Carboniferous rocks north of the Lake District, and the north-western side by the Solway and the county boundary to near the junction of the Liddell and Esk.

West of Carlisle is a patch of low ground occupied by Liassic beds covered by drift.

The Carboniferous rocks, as stated, again come to the surface east of the New Red Sandstone tract, and

Nunnery Walks Waterfall
(*Formed in New Red Sandstone Rocks*)

occupy the Cumberland portion of the Pennine and Cheviot Hills.

Of that part of the county occupied by the slaty rocks, the south-easterly portion bordering on the county boundary line and extending thence north-westward to a line from near Egremont past the heads of Buttermere and Derwentwater to the middle of Ullswater, is formed of the rocks of the volcanic group, save where injected igneous rocks occur as shown on the geological map. Another strip of these volcanic rocks extends from near Penruddock station to Cockermouth. Between these is the great strip of Skiddaw slates, of which outlying patches also occur at Black Combe and along the lower part of Ullswater.

Since the last uplifts forming the dome and the Pennine ridge, the work of rivers and glaciers has largely been concerned in cutting out the valleys, leaving the intervening portions to project as the fells. Much of the work has been done by the rivers, which are able to saw their way downwards, thus deepening the valleys, while rain, frost, and the other agents of the weather cause the material of the valley-sides to be carried downwards to the streams at the base, thus widening the valleys. At a time which as compared with the formation of the rocks which we have described is but as yesterday, though remote as compared with the beginnings of the human history of our land, the district was occupied by masses of ice moving downwards from the upland regions towards the sea, and these masses of moving land-ice produced well-known characteristic marks in the shape of rocks rounded and polished by their action. In addition to this, the ice helped to increase the depth and width of the valleys, and also left much of the material which it ground down and carried away in sheltered spots and lowland tracts to form the stiff clay sometimes mixed with sand, and containing blocks of stone of various sizes, which is known as boulder clay. Here and there the ice left large blocks poised in curious positions on the sides of upland valleys termed

Eskdale Head and Scawfell Pike

(*An eroded valley*)

by geologists 'perched blocks.' The lakes of the district occur in hollows partly due to excavation by this ice and partly to blocking of the valleys by deposits of boulder clay or similar material, some lakes being due entirely to one process, some to the other, and others again to a combination of the two.

Since the glacial period, the action of the weather has caused the upper surfaces of the rocks to be broken up into pieces of various sizes, and parts of the glacial accumulations to be loosened, giving rise to soils. Of these there are four main types, which vary in character according to the nature of the underlying rocks. In the slate tracts, the character of the soil is largely dependent upon the glacial accumulations which have in so many places covered the slaty rocks; where the latter are uncovered by glacial materials, they are often bare of soil. The glacial materials give rise on the whole to a poor stiff stony soil, usually wet, though where much sand occurs in the glacial masses, the soil is looser and drier.

The Mountain Limestone when not covered by glacial materials is usually bare; here and there a short sweet turf occurs. Where glacial accumulations lie thickly over the limestone, the soil naturally resembles that of the slate tracts, but where the glacial materials are thin, a fairly rich soil may be produced.

The third type of soil is formed over the New Red Sandstones. There is often a light sandy loam of a red colour, but on this tract also variations are produced by the presence of glacial materials.

The fourth type is found occupying the sites of former

lakes which have been filled in by gravel and silt, and it also occurs on the estuarine flats. The old lake-sites are scattered over the county. On these flat tracts there has been as a general rule an abundant growth of peat, which yields a rich black soil. With the peat is mixed a variable amount of silt, which with the peaty material causes the soil to be especially valuable.

The Rocking Stone, Thirlmere
(*A perched block*)

5. Surface and General Features.

Cumberland, as stated in Chapter 2, may be separated into three important physical divisions, which must now be considered more fully.

An inspection of the physical and geological maps

inside the covers of the book will show that there is some connection between the geological structure and the height of the ground. If we examine the 600-foot contour-line we shall see that it runs at the foot of the Cheviot and Pennine hills on the east side of the Eden valley, and that on the west side, omitting the minor valleys, it sweeps as a semicircular line north and west of the Lake District hills from Penrith to Millom. Turning now to the geological map it will be found that the ground above the 600-foot contour-line consists of the old slaty rocks with their accompanying injected igneous rocks, and of the Carboniferous Limestone beds, while the ground below that contour-line is occupied by the higher Carboniferous beds (Millstone Grit and Coal Measures), the New Red Sandstone and the Lias. The explanation of this is not altogether simple, but it will suffice to state that on the whole the harder rocks have resisted erosion to a greater extent than those of a softer nature, and therefore tend to stand out as eminences, while the newer and softer rocks have been worn away into lowlands.

Let us consider the characters of these divisions in greater detail.

Taking first the fell tract of the Lake District it will be seen from the physical map, which includes those parts of the Lake District which are in Westmorland and Lancashire (although these counties are left uncoloured), that the district is roughly circular. The centre of the circle is about Scawfell—the highest ground—and from this a number of valleys radiate like spokes from the hub of a wheel. Those in Cumberland, beginning from the south-west and proceeding to the north-east, are the Duddon valley, Eskdale, Miterdale, Wasdale, Ennerdale, and the Buttermere-Crummock, Derwent, Thirlmere and Ullswater valleys. Between these are corresponding spokes of high ground forming ridges between the different valleys. The centre Scawfell Pikes, from

The Summit of Scawfell Pike

which these spokes diverge, has a height of 3200 feet. We may now take the ridges in order from south-west to north-east and notice the chief eminences on each. The highest points on the ridge between the Duddon and Esk valleys are Black Combe 1969 feet, and Harter Fell 2140 feet. This ridge is continued northward to the watershed between the Esk and the Langdale valleys, where it forms the county boundary between Cumberland and Westmorland. On that part of the ridge are Crinkle Crags (2816 feet) and Bowfell (2960 feet).

The ridge between Eskdale and Miterdale rises to no great height, but between the latter dale and Wasdale is Illgill Head, rising to a height of 1978 feet. On its north-western face are situated the well-known "screes" of Wastwater.

Between Wasdale and Ennerdale is much high ground, culminating in the Pillar mountain (2927 feet) and Great Gable (2949 feet), with other eminences not much lower.

A well-marked ridge divides Ennerdale from the Buttermere-Crummock valley. Its highest point, High Stile, reaches 2643 feet.

A wide tract of high fells separates the Buttermere-Crummock valley from that of Derwentwater and Bassenthwaite. The highest point, Grassmoor (2791 feet), lies just east of Crummock lake, but several important hills occur to the east and south-east of this.

East of Derwentwater, and lying between that lake and Thirlmere, is a mountain ridge which attains its highest elevation (1996 feet) in High Seat.

Lastly, between Thirlmere and Ullswater is the Helvellyn range, forming the county boundary for some miles, its highest point, Helvellyn (3118 feet), being on that boundary.

Separated from this radial system of ridge and dale by a tract of comparatively low ground is the roughly circular mass of Skiddaw and the Caldbeck Fells, having Skiddaw (3054 feet) as its highest summit.

The second tract is essentially low ground. It is composed chiefly of New Red Sandstone rocks, except a little strip to the north and north-west of the Lake District, which consists of Coal Measures. In this tract is comprised the plain of Carlisle, extending northward to the Solway; this narrows to the south-east between the Lake District and the Pennine hills, as part of the Vale of Eden with the valley of the Petteril. Between these valleys is a fairly high ridge rising to 937 feet at Penrith Beacon. From the plain of Carlisle a narrow strip of coastal lowland extends between the mountains and the sea to the southern extremity of the county.

The third important tract—the Cumberland part of the Pennine and Cheviot hills—is composed chiefly of Lower Carboniferous rocks which sink gently towards the east. In the south-eastern part of the county these rocks are separated from those of New Red Sandstone age by a great earth-fracture, which comes to the surface at the base of the west side of the Pennine hills, and the resistant rocks of those hills stand out with a steep scarp facing westward, so that the summit-ridge of the hills usually lies one or two miles only from the base. From this ridge the Pennines slope gently eastward into the county of Northumberland.

In part of the Pennine chain between the southern county boundary and the village of Melmerby, a narrow strip of slate rocks lies between the New Red Sandstone of the Eden valley and the Carboniferous rocks of the Pennines. It gives rise to a series of somewhat conical hills in front of the great Pennine scarp. The latter rises into very high ground, usually over 2000 feet above sea-level. The highest hill of this tract is Cross Fell, 2929 feet above sea-level, and situated in an area of continuous high ground far more extensive than that of any continuous tract in the Lake District lying above a similar elevation.

6. The Coast=Line.

From Port Carlisle on the north to Millom on the south the coast-line of Cumberland, leaving out of account minor indentations, has a length of over 70 miles.

Port Carlisle is situated at the seaward end of the estuary of the Eden. From here to St Bees Head the coast forms the south-eastern shore of the Solway Firth. The so-called port was a failure, and has been abandoned. Bowness, to the north-west, is the western termination of the Roman Wall, of which more anon. About four miles south-west of this place is the indentation of Moricambe Bay, near the south-west corner of which is situated a small watering-place, Skinburness, and a very short way south of this is the port and watering-place of Silloth. For twelve miles further the coast is of little interest, having great sand-hills. Then comes the little bay of Allonby, beyond which is the important harbour and town of Maryport, at the mouth of the river Ellen on the northern edge of the Whitehaven coal-field. From here the coal-field borders the coast for another twelve miles to Whitehaven. Between these important towns and ports lies a third, Workington, at the mouth of the river Derwent, and between Workington and Whitehaven is a smaller port, Harrington. All of these ports owe their prosperity and, with the exception of Workington, their very existence, to the coal trade.

Whitehaven Harbour

Three miles south-west of Whitehaven is the important cape St Bees Head, where the coast, which to northward and southward is chiefly flat, rises in high cliffs backed by elevated ground, formed of that division of the New Red Sandstone which takes its name, St Bees Sandstone, from this locality.

From St Bees for twelve miles south-eastward the coast

is flat and very straight, and no harbours are found. The land at the back is low for some miles, and much shingle and blown sand have been formed along the coast. On this part of the coast is Seascale, a rising watering-place with golf-links. Between Seascale and the estuary of the Esk, the sand-hills increase greatly in importance. Drigg Point forms the northern shore of the passage for the

St Bees Head

waters of three rivers, the Irt, Mite, and Esk, and a similar promontory, also of sand-hills, forms the southern shore of the passage. Within the passage the water expands into three estuarine tracts, belonging to the above-named rivers. At low water these show exposed sand-banks with intricate channels. Immediately opposite to the opening communicating with the sea, and standing between the estuaries of the Irt, Mite, and Esk, is Ravenglass. Owing to the silting up of the estuaries, it is of little importance, but in the past it was a port of considerable size, and was used as a port by the Romans.

South of the estuary the coast-line is still low for another twelve miles to Haverigg Point, at the north side of the mouth of the Duddon estuary, though the high ground of Black Combe lies near it between Bootle and Silecroft. This part of the coast is again straight.

Haverigg Point is formed of sand-hills, with alluvium behind. Two miles east of this point is Hodbarrow Point projecting into the estuary, and giving shelter to the indentation of estuary lying northwards, where is the important town and port of Millom, brought into existence by the iron industry. The estuary extends six miles north of Hodbarrow to Duddon Furnace, so-called from the old bloomery or iron-smelting works which still occur though in a ruined state. South of the furnace is the picturesque village of Lady Hall, and east of this the estuary is bridged by the viaduct of the Furness railway. The mid-channel of the Duddon in the estuary forms the county boundary.

Several lighthouses occur along the coast. These are built and supported by a branch of the Civil Service known as Trinity House. The Elder Brethren of Trinity House obtain the funds for the purpose by levying light dues on the ships which enter and leave British ports. The principal Cumbrian lighthouses are two off Silloth, one at Harrington, and one at St Bees Head.

7. Watersheds and Passes.

Cumberland having so much high land, the watersheds have had a most important effect as barriers checking the spread of the people who entered the area from elsewhere at various times, while on the contrary the passes have allowed them to pierce these barriers, and gain access to lower ground on their further sides. It is necessary therefore that we should consider these barriers and their passes in some detail, and that they should be studied on the map.

There are two important watersheds in Cumberland and Westmorland which are connected with one another so as to form a rough letter T laid on its side thus ⊢. Of these the part of the T on the east is the Pennine chain which here forms the watershed dividing the rivers which flow eastward into the North Sea from those which flow westward into the Irish Sea. The other part separates the rivers of the Lake District which flow northward and north-westward into the Solway and that part of the Irish Sea which lies north of St Bees Head from those which flow southward into Morecambe Bay.

The two lines forming the T meet in Westmorland at the head of the Eden valley, but the northern part of the Pennine watershed and the western part of that of the Lake District are in Cumberland.

The low grounds of Cumberland and Westmorland, which on account of their comparative fertility were the tracts desirable to the early invaders of the territory, lie

in the angles where the two lines forming the **T** join.
The northern angle is occupied by the lowlands of the
Eden valley, which extend also over the Carboniferous
rocks south of the Eamont valley to the foot of Ullswater,
and continue northward to form the plain of Carlisle.
The lowlands of the southern angle are in Westmorland.

The low grounds of the county could be readily
reached by three routes : (1) from the sea; (2) from the
south of Scotland; (3) through a pass between Carlisle
and Haltwistle, forming a way between the Tyne and
Eden valleys. This pass, separating the Cheviots from
the Pennines, is the principal Cumbrian pass to the east,
but there are two passes over the Pennines which are of
importance, though only one is in Cumberland. This is
the pass over which the great road made by McAdam
runs from Penrith to Alston Moor. Its importance was
due to the lead industry of the Alston district. It attains
a height of 1900 feet at the summit—Hartside Cross.
The other pass, Stainmore (about 1400 feet high), is in
Westmorland, but is of importance to us, as it forms a
way between the east of England and the low ground at
the head of the Eden valley which is continuous with
that of the lower part of the valley and of the Cumberland plain.

The principal passes in the Cumberland portion of
the second branch of the ⊣, which is the main watershed
of the Lake District, may now be noticed. Beginning at
the east, the first is the important pass of Dunmail Raise,
which is only 783 feet above the sea, and has at all times
formed a route between the north and south of the Lake

District. About five miles south-west of this is the Stake Pass (1581 feet) between Langdale and Borrowdale. A minor pass, Rossett Gill (2106 feet), lies less than two miles east of this, on the upland route between Langdale and Wasdale, though it really separates the former from another tributary of Borrowdale. These three passes, which are mainly important to tourists, lie on the county

Wasdale Head, showing Pass into Borrowdale

boundary between Cumberland and Westmorland. About one mile westward from Rossett Gill is Esk Hause (2420 feet) between Borrowdale and Eskdale. It is the highest important pass in the county and is of the utmost import to tourists. Two miles north-west of Esk Hause is Sty Head Pass, connecting Borrowdale with Wasdale, and at a similar distance from this proceeding in the same direction is Black Sail (1800 feet) between Wasdale and Ennerdale.

North of the main Lake District watershedding line several passes indent the ridges between the important valleys of Ennerdale, Buttermere, Derwentwater, Thirlmere, and Helvellyn, and to the south is the important

Honister Pass and Crag

Hardknott Pass (1200 feet) between the Duddon and Esk valleys. Three of the northern passes are occupied by carriage roads, namely Honister (1190 feet), Buttermere Hause (1096 feet), and Whinlatter (1040), all lying between the Buttermere and Derwentwater valleys.

One important pass remains to be considered, which separates the Skiddaw group of fells on the north from those of the Helvellyn mass on the south. This is near Troutbeck station, at a height of nearly 1000 feet above the sea.

There are, of course, many minor watersheds separating vale from vale, with passes over them, but they have played little part in the distribution of the population, and require no notice.

8. Rivers.

The principal river of Cumberland is the most easterly stream, the Eden, which rises in the Pennine Hills some ten miles south of Kirkby Stephen, and enters the county a few miles east of Penrith. The other important rivers rise in the centre of the Lake District dome in the Scawfell group of hills, and flow to the north, north-west, west, south-west and south. The first of these is the Derwent, and proceeding westward and southward we meet with the Ehen, the Irt, the Esk, and the Duddon. In addition to these are some minor streams, the Wampool, the Ellen, the Calder, and the Mite, which are exceeded in size by some of the tributaries of the larger rivers.

The Eden throughout its course through Cumberland flows over ground formed of New Red Sandstone rocks. Its Cumbrian waters extend through about 35 miles of country, the total length of the river being over 60 miles. Flowing at first entirely through Westmorland, it forms the county boundary for a short distance, but after receiving the Eamont from the west flows through Cumbrian territory to the Solway. The Eamont flows from Ullswater, and from its furthest tributary near Kirkstone Pass in Westmorland to its junction with the Eden has a length of about 30 miles. At Kirkoswald the Eden, which has hitherto occupied a wide valley, enters a picturesque gorge, which continues to Armathwaite, beyond which the valley expands and passes gradually into the plain of Carlisle. Several minor streams from the Pennines flow into the Eden between Langwathby and Wetheral. At Newby, east of Carlisle, the Irthing, rising far to the north-east on Sighty Crag in the Cheviots, joins the Eden, and at Carlisle, first the Petteril from Penrith, and then the Caldew from the Caldbeck Fells come in. The border-stream, the Esk, with its tributaries might perhaps be regarded as distinct from the Eden, inasmuch as it has a small estuary separated from that of the Eden by a narrow tract of marshland, but as the two rivers enter the Solway close together they may be regarded as components of one system.

Several streams rise on the north side of the Caldbeck Fells and their waters, entering the Wampool and Waver, are discharged into Moricambe Bay. The Ellen also rises on the Caldbeck Fells, but at their western end, and after a course of about 20 miles enters the sea at Maryport. South of this, the Cumbrian river which is second in importance—the Derwent—falls into the sea at Workington. It rises on the Scawfell group, and has a total length of between 30 and 40 miles. The actual source is usually placed in Sprinkling Tarn below Great End,

The Borrowdale Birches

whence it flows northward to Rosthwaite, where it is joined by the Langstrath Beck, and the united streams continue to Derwentwater. From its source to Derwentwater the valley of the Derwent is known as Borrowdale. In the alluvial flat between Derwentwater and Bassenthwaite it receives the Greta from the east (draining the Thirlmere valley and the valley from Troutbeck station),

Birks Bridge over the Duddon

and Newlands Beck from the south-west. Leaving Bassenthwaite the river runs on the whole westerly to its mouth, receiving one important tributary, the Cocker (which comes from Buttermere, Crummock, and Loweswater through the Vale of Lorton), at Cockermouth.

The Ehen, called in its upper course the Liza, rises on Great Gable. About six miles from its source it enters

Ennerdale lake. Flowing from that lake in a general westerly direction for a similar distance it reaches Cleator, turns south, and after yet another six miles enters the sea at Sellafield.

The Calder rising in Copeland Forest, after a course of less than 10 miles, discharges its waters close to the mouth of the Ehen.

The three-branched estuary at Ravenglass receives the waters of the Irt on the north, the Mite in the centre, and the Esk on the south. The Irt flows from Wastwater, and receiving the Bleng from the north, proceeds by a tortuous course to its estuary. The Mite rises on Scawfell, and has a short straight course. The Esk rises at Esk Hause, and flows through Eskdale, the total length of the valley approaching 20 miles.

Lastly, the Duddon, rising at the Three Shire Stones, where Cumberland, Westmorland, and Lancashire join, forms the boundary between Cumberland and Lancashire from source to sea, a distance of about 15 miles.

9. Lakes.

Cumberland possesses many lakes and tarns. The term tarn as a whole is applied to a small lake, usually less than half-a-mile in length. The larger lakes occupy parts of the floors of the valleys, while the greater number of the tarns are perched in hollow combes on the hill sides far above the valley bottoms, and in many cases the streams which come from them flow in cascades down the sides of the larger valleys.

We will consider first the lakes according to the valleys in which they lie, starting in the east, and working round by north to west and south-west.

Ullswater has its upper reach entirely in Westmorland, while the county boundary runs along the centre of the middle and lower reaches, so that their north-westerly parts lie in Cumberland.

Thirlmere at the present day

In the Derwent drainage are three lakes. Thirlmere is in an eastern tributary valley, while the main valley has Derwentwater and Bassenthwaite, once forming one lake, but now separated by an alluvial flat. In the same way, a flat separates two lakes formerly united, which lie in the next important valley to the west ; these lakes are Buttermere and Crummock. Loweswater is in a tributary valley which joins the main valley at the foot of Crum-mock. Ennerdale lies in the Ehen valley and Wastwater in that of the Irt.

Let us now consider the sizes and depths of the lakes.

Ullswater is about $7\frac{1}{3}$ miles long measured along the central line of the lake, and has an area of nearly $3\frac{1}{2}$ square miles, its greatest breadth is 1100 yards and the

Thirlmere: before the Enlargement of the Lake

average breadth over 800 yards. It is 476 feet above sea-level. The greatest depth is 205 feet, at a point about $1\frac{3}{4}$ miles from the head. The lake extends on the whole from south-west to north-east but is far from straight, having a rude Z-shaped form. The upper or southern reach is the shortest, being about one mile long, and stretching from south to north. The other reaches are each about three miles in length, the middle reach lying about west-south-west—east-north-east, and the northern more nearly south-west—north-east. There is one important bay, Howtown Bay, at the head of the northern reach. The Glenridding Beck has built a large delta near the lakehead, and Sandwick Beck another on the southern side of the middle reach. A few small rocky islets occur in the upper reach.

Buttermere, Crummock, and Loweswater from Honister Crag

Thirlmere was about 2½ miles long and less than half a mile wide, with a depth of over 100 feet. It was nearly divided into two at a narrow neck, with a picturesque bridge. It has been artificially enlarged and deepened, and now forms the reservoir for the water supply of Manchester.

Derwentwater is nearly three miles long, and 1⅕ mile wide in its widest part. It has an area of about two square miles. It lies 245 feet above the sea. Its greatest depth is 72 feet, so that compared with the other lakes of the district it and Bassenthwaite are exceptionally shallow. It has several small bays, and a number of islands formed of glacial deposits, and not of solid rock.

Ennerdale Lake

Bassenthwaite is nearly four miles long and three-quarters of a mile wide at its widest part. Its area is the same as that of Derwentwater. It lies 223 feet above sea-level. Its greatest depth is 70 feet. It has several bays and one or two small islets. It was at one time much larger, and formed one sheet of water with Derwentwater, as already stated.

M. C.   4

Buttermere is 1¼ mile long, with an average breadth of 620 yards and an area of over one-third of a mile. Its greatest depth is 94 feet. The sides are remarkably straight. It lies about 330 feet above the sea. Crummock Water has a length of 2½ miles, a mean breadth of about 700 yards and an area of nearly one square mile. It has in one place a depth of 144 feet. It is about 320 feet

Wastwater

above sea-level. Near its south end is Scale Force (125 feet).

Loweswater is about 1¼ mile long, with a maximum width of about 600 yards. It is said to be 60 feet deep, but no accurate survey has been made. It is about 430 feet above the sea.

Ennerdale Water has a length of nearly 2½ miles, and an average breadth of 800 yards, with an area of over one square mile. It is nearly 370 feet above sea-level. Its greatest depth is 148 feet and is near its foot.

Wastwater is three miles long with an average width of 650 yards, and an area of more than one square mile. It is about 200 feet above sea-level. The deepest point is 258 feet, so that it lies 58 feet below sea-level and about 200 acres of the floor lie below that level. It is the deepest lake of Lakeland, and the only Cumbrian lake which has part of its floor below sea-level.

A considerable number of tarns are found in the uplands. We will mention the principal, arranged according to the valleys to which they belong. In the Derwent valley, Angle, Sprinkling, and Sty Head tarns lie on the north side of the Scawfell group at the head of the valley, and Blea and Watendlath tarns in the Watendlath valley, the stream of which flows over Lodore Falls into Derwentwater. Bleaberry and Floutern tarns are in the drainage of the Cocker; Scoat, Low, and Greendale tarns in that of the Mite, and Burnmoor and Devoke in that of the Esk. The latter is the largest tarn in the district, and to some extent forms a connecting link between the tarns and the valley lakes.

10. Scenery.

In a county of which the scenery is an important factor affecting the inhabitants, some attention must be paid to the causes and character of that scenery. These are dependent partly upon the geological structure, partly

upon meteorological conditions whether acting directly—
(e.g. the effects of sunlight and clouds upon the view)—or
indirectly—(as affecting the vegetation), and partly also
upon the agents such as frost, rivers, and glaciers, by which
the details of the scenery have been largely determined.

At the outset we may take into account the effects of
the more important rock-groups in controlling the nature
of the surface.

Sprinkling Tarn

Beginning with the slate-rocks, the oldest beds, the
Skiddaw slates, are somewhat easily worn away, and with
no great variety of rocks, give rise to conical hills with
grass-covered slopes and few rocky cliffs. Skiddaw is
a good example of this type of hill.

The rocks of the middle division of the slates, belonging to the Borrowdale volcanic group, are responsible

for the wildest scenery in the county. They are the hardest of the rocks which are extensively found, and as there is considerable variety in their hardness, the lavas and many of the volcanic ashes being peculiarly hard, while some of the ashes are softer, a considerable diversity of outline is thus caused by these alternations. Again, they are affected by very regular systems of gigantic cracks or joints, often with belts of smashed rock along the cracks, and these cracks have been lines of weakness which have frequently been worn into notches and gorges, and they also define the sides of cliffs, and of rock pinnacles. To the hardness of the rocks we owe in a great degree the superior elevation of the fells which are composed of them ; and to the variations of hardness and the nature of the joints, the frequent alternations of cliff and slope which are so marked a feature of these fells, and are displayed at their best in the Scawfell group of hills. On some of the more level ridges this type of scenery is replaced by smoother outlines, due to the abundant growth of peat.

The upper slates give rise to tamer scenery, but the rocks are barely represented in Cumberland.

The Carboniferous rocks are chiefly limestone (though above the limestone we find in some places Millstone Grit and Coal Measures), and in this group is the mass of Whin Sill noticed in the geological chapter.

The structure which is presented by the Mountain Limestone hills has well been named " writing-desk structure," for the gently inclined beds form gentle slopes, with steep cliffs determined by the nearly vertical joints on the other sides of the hills. This structure is typically

The Pillar Rock, Ennerdale. (*Formed of Volcanic Rocks*)

shown in the limestone tract between Ullswater and Hesket Newmarket and in the Pennine chain east of the Eden valley. The bare white cliffs and fissured " clints " are a marked contrast to the surfaces of the fells formed out of the slate-rocks.

The clints just mentioned are flat or fairly flat surfaces of limestone with fissures produced by the widening of the vertical joints by acidulated rain-water which has extracted carbon dioxide from the atmosphere. This water is capable of dissolving the limestone, and the bare flat or gently-sloping limestone surfaces are therefore often traversed by two sets of fissures at right angles to each other, penetrating for many feet or even yards from the surface. The sides of these are often honeycombed by the solvent action of the rain, assisted by the vegetation that may grow abundantly within.

The massive well-jointed Millstone Grit, when found on fell-tops, produces flat-topped hills with steep scarps or cliffs beneath, well shown on the summit of Crossfell.

The rocks of New Red Sandstone age, being soft, are readily worn away, hence the comparatively low ground occupied by them in the Eden valley and the north Cumbrian plain. Furthermore, their rapid breaking-up allows abundant formation of soil, and the bare rock is seldom naturally exposed save along the sides of some gorges. In addition to this a great deal of glacial material has accumulated in the low ground of the Eden valley and has masked the rocks beneath.

The effect of the ice of past ages in helping the hollowing of the valleys has been noticed in the geological chapter. We are here concerned with its deposits. We may first notice the little moraines which were left by the upland glaciers in the more central valleys as piles of rubbish at their ends. They consist of hummocky hillocks of clay, gravel, and stones, covered with coarse vegetation, and produce a somewhat desolate effect. These moraines may be well seen in several of the upland valleys such as Borrowdale. A large number of the tarns are partly held up by moraines.

In addition to these moraines, the boulder-clay which was noticed in the geological section is spread widely over the low-lying grounds. It is often arranged so that its upper surface forms parallel mounds like the backs of whales. These are known as "drumlins."

Latest of the deposits which have produced large effects upon the scenery are those which fill in lakes and the estuarine tracts. Long flats occur at the heads of the principal lakes, where the rivers are pushing their deltas forward and converting water into land. Scores of old lakes scattered over the county have thus been filled, and the upper surface is usually occupied by peat growth. The accumulation of frost-riven blocks on the slopes to form "screes" should also be noticed.

Towards the coast, the scenery is often due to accumulation of deposits in recent geological times. Along the shores of the Solway and at the heads of some other estuaries we find peat-mosses, due to growth of vegetation upon silt laid down by the waters of these estuaries. Such is Solway Moss.

Nearer to the waters of the estuaries we often come

The Screees, Wastwater

across salt-marshes, while still lower are sand-banks covered by tide at high-water.

From Skinburness to Allonby Bay, also on either side of the estuary of the Irt, Mite and Esk, and again at the extreme south end of the county, are sand-dunes caused by the piling up of sea-sand by on-shore winds. They are often partly covered with a growth of spear-leaved vegetation, which to some extent binds the sand-grains together.

The waterfalls of Cumberland are remarkable for their beauty rather than their size. Many occur at the junction between hard and soft rocks, and have usually cut gorges some way back into the hard rock. An example of such is Scale Force. Others are cascades where the waters from the upland valleys pour down into the main stream, as Sourmilk Gill and Taylor Gill near Seathwaite. Lodore unites the characters of each of these types.

The influence of vegetation on the scenery will be noticed in the chapter treating of Natural History, and as for the atmospheric effects, one need only remark that the variability of the climate which is sometimes treated as a matter of regret is responsible for scenic effects which are far more beautiful than would be the case were the climatic conditions of a more settled character.

11. Natural History.

Botany and zoology are the sciences which treat of the world's flora and fauna, but the study of the distribution of plants and animals—where they are found and why—forms part of the domain of geography, for from these we learn many facts concerning the past history of the land. Every one knows by sight a certain number of the plants and animals of his own county, and this knowledge will enable him to get some idea of the way in which their geographical distribution is effected.

Let us in the first place consider the plants of the county. Some of these are commonest in the south of England, others in the east, and others again in Scotland, while a very large number of the whole are spread over the entire island, and a few are very local, so far as our country is concerned.

These plants have not originated where they now grow. We have seen in the geological chapter that the district was once occupied largely by ice. At that time a few plants may have lived on the rocks of the higher fells, just as they do on the hills appearing above the ice of Greenland at the present day. But as the ice receded from the county tracts must have been left bare on which plants gradually sprang up, as their seeds were wafted by the wind, or brought by birds, or in some other manner, from other regions.

It must not be supposed that the plants which were thus brought came from the regions above mentioned where they are now commonest, any more than that the Celtic speaking people of Britain who are now confined to the western high grounds entered from the west. There are many reasons for believing that at no remote geological date, though before the beginning of historic times, England was joined to the Continent along the tract now occupied by the Straits of Dover. This would form a ready route along which the plants which undoubtedly reached England from the Continent could gradually migrate, just as, at a later period, successive immigrations of people came along that route, having only to cross the narrow straits. And as the more barbarous people were driven into the mountain fastnesses by their more highly civilised successors, so might the early plants be replaced by others which, under altered climatic conditions, were able to flourish.

But not all the human immigrants into Britain came by way of the narrow passage of the Straits of Dover. The sea-faring Danes and Norsemen, for instance, landed sometimes on the north-east and even on the north-west coast of England. Similarly some of our plants may have come in along some other route when England was united to the Continent not merely along the tract now occupied by the straits, but by land masses which once existed over part of the site of the North Sea.

Certain plants now well established have been introduced by man. Most noticeable among these are such as grow in cornfields which have been accidentally brought into the country with the corn. Some have been recently introduced and are not yet established, as the small toad-flax ; others, like the large blue speedwell (*Veronica Buxbaumii*), though of recent introduction, are now thoroughly established; and some, like the blue cornflower, have been so long inhabitants of the country that the period of their introduction is unknown.

It will be seen from these remarks that the question

how our county became stocked with its plants is very complicated, and as it requires much knowledge of science to sift the evidence, this part of the study of distribution is not for anyone who is not possessed of much botanical knowledge.

There are other facts connected with distribution, however, which can readily be tested. It will soon be found that the plants of Cumberland do not flourish equally in all parts of the county, for instance, those growing on the flats near the sea are very different from those which live on the slopes of the higher parts of Scawfell and Helvellyn. Very little observation will show that there are two important causes of this difference among the plants of various tracts of the county, namely height above sea-level, and difference of soil. Let us first regard the difference of altitude. Many plants are confined to tracts less than 900 feet above sea-level, of which the gorse or whin is a good example. Above 1800 feet the bracken practically ceases. In the belt between 1800 and 2700 feet we find a remarkable assemblage of plants of an alpine character, such as the kidney-leaved sorrel (*Oxyria reniformis*), the rose-root (*Sedum rhodiola*), the purple-flowered saxifrage (*Saxifraga oppositifolia*), and the alpine rue (*Thalictrum alpinum*). Finally, above the last-mentioned height we get the most arctic of all the Cumberland plants—a little creeping willow (*Salix herbacea*) which is characteristic of the hills above 2700 feet. It will be a useful exercise for the student to discover for himself what are the upper limits of the various plants with which he is acquainted, though he must be prepared to find an occasional straggler above the height to which the species as a whole ascends.

An easier study is that of the distribution of plants according to the soil, it being remembered that this soil in many cases varies in character according to the nature of the rock beneath, though some soils, like those formed of peat, are largely independent of the underlying rock.

Some plants are confined to the muddy silt of the salt-marshes by the sea-shore. A conspicuous example is the purple sea-aster or starwort with yellow eye (*Aster Tripolium*), which grows on the salt-marshes of the estuaries. Others as the sea-bindweed (*Calystegia soldanella*) and the yellow-horned poppy (*Glaucium luteum*), are rooted in the coastal sand-dunes.

The bog-plants live in bogs at various heights from sea-level to the tops of some of the highest hills. Near sea-level we may yet meet with the Osmunda or royal fern, though it is less common than formerly owing to ruthless destruction by collectors. Even yet it is in places sufficiently plentiful for use as litter for cattle. Other interesting bog-plants are the louseworts, the insect-eating plants known as butterwort and sundew, the beautiful little mealy primrose—the "bonny bird e'en" of Lakeland (*Primula farinosa*)—and, handsomest perhaps of all, the grass of Parnassus (*Parnassia palustris*). In the pools among the bogs we find other plants as the bladderworts (*Utricularia*) and the pale blue-flowered water lobelia (*L. Dortmanni*).

Some plants are confined to the rich soils along beck sides like the globe-flower (*Trollius europaeus*) and the yellow balsam (*Impatiens noli-me-tangere*), the latter a truly local plant.

In the rough pastures we may observe many kinds of orchis, and along the hedgerows one often comes across two conspicuous plants of Scotch type, namely the melancholy thistle (*Carduus heterophyllus*), and the great bell-flower (*Campanula latifolia*).

Borrowdale Yews and Seathwaite

In the limestone district is a group of plants which flourish notwithstanding the general dryness of this tract. Among such are the centaury, the rock-rose, and the lady's finger. But the most noticeable plants of the limestone tracts are those which live in the fissures of the cliffs and clints, such as the harts-tongue and green spleenwort ferns, and the yew. The last named also flourishes on the volcanic rocks which contain some lime.

The determination of plants growing on different kinds of soil will be a useful exercise. It will sometimes be found that the same plant has a different growth under different conditions ; thus, the golden-rod of the lowlands is not quite like that found in the fells, and the vernal whitlow-grass (*Draba verna*) has a different growth on limestone cliffs from that which it exhibits on the slate-rocks.

Before leaving the consideration of the distribution of plants there is one matter concerning which a few words must be said.

We saw that at heights of over 1800 feet a number of plants were found which do not occur below this altitude. These however are widely scattered in European mountain regions, many of them being found on the Scotch hills, the Alps of Switzerland, the mountains of Norway, and on lower ground within the arctic circle. They are at the present day mainly characteristic of alpine and arctic regions, and it is believed that they became established in our country during the glacial period, occupying then the British lowlands, just as they now live on the lowlands within the arctic circle. As the climate improved they were there displaced by other plants which were able to flourish to so great an extent as to exclude these "alpines," which accordingly were driven higher and higher, and are now found obtaining here and there a precarious footing upon our higher fells, from which perhaps they are doomed to disappear at no distant date. Let us hope that the disappearance, if it comes, will be natural, and not quickened by the wanton removal of the roots of the plants by the too eager collector.

In a county of which the scenery has within recent times had a marked effect upon the dwellers therein, a few words may be added as to the effect of the plants upon that scenery.

Many plants grow in sufficient number to produce a striking influence upon the landscape. The flowers of the rag-wort in the rough pastures, the curious growth of the cotton-grass when in seed, and the leaves of the alpine lady's mantle, which, growing in masses on the rock ledges of the fell-sides sometimes gleam with a green of almost metallic sheen, may be cited as examples. There are, however, two plants whose influence is particularly pronounced, namely the heather and the bracken. The effects of heather are most striking on the moors beyond the true Lake District, the amount of heather in the Lakeland portion of Cumberland being comparatively small, but it is in this tract that the effects of the bracken are so fine. Of it Wordsworth speaks thus:—"About the first week in October the rich green which prevailed through the whole summer is usually passed away. The brilliant and various colours of the fern are then in harmony with the autumnal woods : bright yellow or lemon colour, at the base of the mountains, melting gradually, through orange, to a dark russet brown towards the summits, where the plant, being more exposed to the weather, is in a more advanced state of decay."

About the animals of the district we can say little. Gifted with power of locomotion, their distribution is as a whole wider than that of the plants.

The mammals have suffered much from the hands of

man, and especially is this the case with the larger beasts. The wild cat, the wild boar, and the badger are now extinct, though their previous occurrence is indicated by some of the place-names. There are several Grizedales in the Lake District. These dales are so named from "gris," an old name for the wild boar. Again, the word Brockstone means "the stone of the brock" or badger. The fox is yet with us, and a fell fox-hunt is still an exhilarating pursuit, conducted on foot. The otter frequents the streams and it also is hunted.

Birds are abundant enough in the lowlands, though on the fells there is a singular lack of bird-notes. The call of the wheatear is an exception, and we may often hear the cry of one or other of the species of hawk, and more rarely the hoarse croak of the raven. The grouse-moors of the county lie chiefly outside the actual Lake District. The red grouse is confined to Britain. Sea-fowl are abundant on the coast, especially in the estuaries. Near Ravenglass, a breeding-place for sea-birds is pro-tected : here large numbers of gulls, terns and other birds have their nests.

There is little of general interest with regard to the distribution of the reptiles and amphibians, but the fishes of the county present some noteworthy features. Char are found in Ullswater, Buttermere, Crummock, Ennerdale and Wastwater, and in Ullswater is the fish known as the schelly. Both char and schelly in Britain are characteristic of the lakes of hill-regions. So far back as the beginning of the eighteenth century we learn that char were "baked in pots, well seasoned with spices, and sent up to London as a great rarity," and a few still find their way thither.

Of the invertebrate creatures much could be written in detail, but it would require considerable knowledge of zoology. Leaving the mass of these animals unnoticed we may refer to one, the mountain ringlet butterfly (*Erebia cassiope*), which is abundant at a height of over two thousand feet on the fells, and cannot be met with between here and Switzerland. Like the Alpine plants it is probably a survival from the organisms which spread over our country during the Glacial Period.

12. Climate.

The climate of a country is the result of the combined effect of the different variations of what is commonly termed the weather. The most important factors in determining the climate are temperature and rainfall.

The great variations in the climate of the world depend mainly upon differences of latitude; thus we speak of tropical, temperate, and arctic climates; that of our country being temperate. Another important factor in controlling climate over wide tracts of country is nearness to the sea, so that along any great climatic belt we have variations according to the geographical conditions, the extremes being "continental climates" in the centres of continents far from the oceans, and "insular climates" in tracts surrounded by ocean. The continental climates are marked by great variations in the seasonal temperatures, the winters tending to be exceptionally cold and

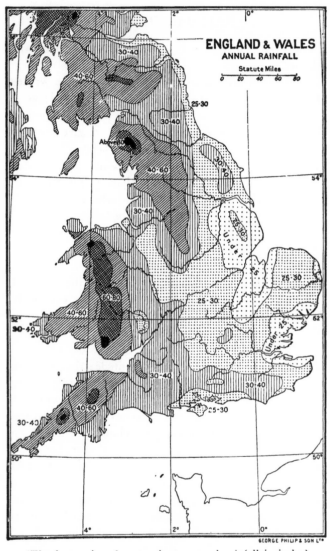

(The figures give the approximate annual rainfall in inches.)

the summers exceptionally warm, whereas the climate of many insular tracts, including Britain, is characterised by equableness,—by mild winters and fairly cool summers. Again, an insular climate tends to be more humid than a continental climate. Great Britain, then, possesses a temperate insular climate.

Different parts of England possess different climates, and we must now consider wherein and why the climate of Cumberland varies from that of other parts of the country.

Two especially important points must be regarded in contrasting the climatic conditions of Cumberland with those of other parts of England. Firstly, Cumberland is further from the European continent and nearer to the Atlantic Ocean than is the eastern portion of England, and its climate therefore departs more widely from the continental type than does that of eastern England. In the second place the Cumberland climate is largely influenced by the great amount of elevated land within the county boundaries.

As the evaporation of water and its subsequent precipitation as rain is dependent upon changes of temperature, we may consider first the temperature changes.

England and Wales are situated in a belt having a mean annual temperature of about 50° Fahr., the mean temperature for January being about 40° Fahr., and that for July 60° Fahr., and these figures hold good for Cumberland, whereas in East Anglia the January and July temperatures are about 38° and 62° and in parts of western Ireland about 42° and 58° respectively. It will be seen then that, comparing summer and winter temperatures, East Anglia has a less equable, and western Ireland a more equable, climate than Cumberland.

This distribution of temperature shows that latitude alone does not produce the variations, otherwise it should be colder as one passes northward. It has long been known that temperature variations in our island are greatly affected by the prevalent south-westerly winds bringing heat from the waters of the Atlantic. These waters off our coasts are exceptionally warm for· their latitude, owing to their movement from the warmer south-westerly seas towards our shores on the north-east. This movement is that of the Gulf Stream, a drift of the surface-waters of the Atlantic in a north-easterly direction caused by the prevalent winds.

It is impossible here to discuss the principles which control weather changes. It must suffice to say that our weather is largely influenced by the prevalence of *cyclones* from the Atlantic. The air movements are *cyclonic* or *anticyclonic*. In a flowing stream we may often observe a chain of eddies bounded on either side by more gently moving water. Regarding the general north-easterly moving air from the Atlantic as such a stream, a chain of eddies may be developed in a belt parallel with its general line of movement. This belt of eddies, or cyclones, as they are termed, tends to shift its position, sometimes passing over our islands, at others to the northwards and at others again to the southwards. To the shifting of this belt most of our weather changes are due.

When the country is influenced by a cyclone it is

often windy, while when under the influence of an anti-cyclone it will more probably be still and dry. Cyclones, then, are apt to be accompanied by wind and rain, anti-cyclones by calm, during which there may be bright sunshine with warmth in summer, clear cold weather in winter, and fog in autumn.

There is one period of the year when the distribution of the winds in our county is affected in a different way by the temperature of the great continental mass to the east. The conditions are then such that the belt of cyclones is, as it were, pushed back over the ocean, and we experience the east winds which are often prevalent during the month of March.

Let us now further consider the rainfall. Cold air can hold less water vapour than hot air, and accordingly when the air rises and becomes chilled in the higher parts of the atmosphere it tends to part with its moisture as rain. This air may rise by expansion, which makes it lighter, or by blowing up a rising land-surface. The importance of the latter cause is great, as may be seen by studying a map of rain-distribution in our island, when it will be noticed that the areas of high rainfall coincide with the elevated tracts. The large amount of rain which falls in Cumberland is mainly due to the vapour-laden winds from the Atlantic being forced up the hills and precipitating their moisture, and accordingly the greatest amount of rainfall occurs practically on the tops of the ridges which face the ocean. The greatest rainfall in the county occurs in the westerly tract of high ground, where it is over 80 inches per annum. It decreases eastward, and sinks to under 40 inches in the Eden valley. The rainfall of the great Pennine ridge east of that valley has not been studied in detail. It may be remarked by contrast that the driest part of England has less than 20 inches of rain per annum.

The figures given as to the annual rainfall in the Lake District in some books are very misleading, and Seathwaite in Borrowdale has an unenviable reputation for an average rainfall of 154 inches per annum measured in six years. But this is exceptional. It must be remembered also that the amount of rainfall does not give a measure of the length of the period during which it falls. The very heavy falls in parts of Cumberland cause a much greater amount to be precipitated in a given time than in some other districts where the fall is gentler.

The amount of sunshine recorded differs for different parts of our island, the greatest amount on the whole being in the south, and the least in the southern part of the Pennines. Along the greater part of the south coast more than 1700 hours per annum are recorded. The smallest amount for England is under 1200 hours. Parts of Cumberland have less than 1300, and no part has more than 1500 hours.

From the foregoing remarks it will be gathered that the bad reputation of the district is not altogether deserved, and in the summer months, during which visitors arrive in greatest number, the number of hours of rainfall is not excessive.

One or two matters of detail in connection with the Cumberland climate should be noticed.

As regards wind the valleys of the upland tracts are comparatively sheltered and violent gales are not frequent, though the wind is often very strong on the ridges. One local phenomenon deserves mention, namely the "Helm wind,"—so called because it is accompanied by a helmet-shaped cap of cloud on the Cross Fell ridge. It is an east wind, and the phenomenon is most pronounced during the prevalence of the early spring east winds. Leaving subsidiary features out of account, the wind does not differ from winds in other parts of the world which have a similar configuration. The Pennine ridge, as we have seen, has a gentle slope eastward and a steep scarp facing west. The air rises slowly up the gentle eastern slope and rushes violently down the western scarp, and its moisture becomes condensed on the summit ridge.

Severe frost is not so frequent in Cumberland as it is in parts of south-eastern England, where the average winter temperature is lower. Snow falls in the winter season on the higher fells and often lies long there, but there is no very great amount of snow in the lower regions. Some of the most severe snow-falls occur during the prevalence of east winds, and therefore tend to be heavier on the Pennine ridge than upon the hills of the Lake District.

13. People—Race. Language. Settlements. Population.

When the Romans invaded our island, it was occupied by people whom we are accustomed to speak of popularly as the early Britons. These people, however, were not all of one race, and we may briefly consider who they were. Of the earliest inhabitants of our land, known as the "Palaeolithic" men, we have no trace save the implements which they left behind, and of these none have been found in Cumberland. Long after the disappearance of these people, a short swarthy race arrived from the Continent and spread widely over Britain, certainly penetrating into Cumberland, as indicated by their relics. Whence they came and who they were we know not for certain; all we can say is that they were an earlier set of immigrants than the Celts who succeeded them.

These early men were displaced by a taller and more powerful people armed with better weapons, who, however, probably did not completely destroy their conquered enemies, but held the survivors in bondage as slaves. The more powerful race, the Celts, are supposed to have come into Britain at two distinct times. The earlier immigration was of a Celtic race who spoke a language like the modern Gaelic: these people are known as the Goidels or Gaels. Subsequently another Celtic race—the Brythons—speaking a language like Welsh arrived.

There were, then, in our county, even if palaeolithic man never arrived there, three races before the Roman invasion, one pre-Celtic and two Celtic, though some believe that the inhabitants of what is now Cumberland were essentially Goidels. Be this as it may, at the time of the arrival of the Romans, the north of England, including Cumberland, was occupied by a powerful Celtic tribe, that of the Brigantes. This tribe was divided into sub-tribes, but their distribution is uncertain, and it will be

sufficient for our purpose to know that the people of the
Brigantes in those days inhabited what is now Cumberland.   It is doubtful whether any traces of the Brigantes
can be found among the characteristics of the existing
people of Cumberland.   But many of their place-names
still survive.

In the first century A.D. the influence of the Romans
began to be felt, and was exerted in Cumberland for
nearly four hundred years.   Important as was the civilising
influence of the Romans upon the inhabitants, as we shall
see in later chapters when we come to speak of the roads
and other relics of that people, the Roman occupation produced little permanent effect upon the physical characters
and the language of the inhabitants.   The occupation was
essentially military, and the Roman legions were composed
of a soldiery of mixed race gathered by the Romans from
various quarters of Europe.   During the temporary wane
of Roman influence in the fourth century and after the
final independence of Britain from the Roman yoke in
the following century there were invasions of the district
from the north by the Picts and Scots, but these appear
to have been of the nature of raids, and to have had little
effect upon the character of the modern inhabitants.

In the seventh century the Anglo-Saxons entered the
district by the low ground between the Pennines and the
Cheviots, and towards the end of that century the district
around Carlisle appears to have been incorporated as part
of the Saxon kingdom of Northumbria.   These immigrants were Angles, but a few Saxons may have entered
Cumbrian territory from the south.

In the ninth century the Danes entered the district from the east over the passes of the high lands, and in the tenth century the Norsemen, who had previously invaded the Isle of Man, came over sea to the fertile tracts along the western coast, and settled permanently in the district.

William Rufus, in the year 1092, brought an army to the north and the Norman settlement began. This was the last important immigration of the various races into what is now Cumberland. Of these invasions that of the Romans had a striking effect upon the civilisation of the people ; the Anglo-Saxon invasions of England gave us our language (afterwards modified by Norman influence); while the present physical characters of the inhabitants of Cumberland are with good reason considered to be chiefly due to the incoming of the Danes and Norsemen.

It has been well said that the history of a country is written on the face of the country itself—in the names of its towns and villages, its rivers, mountains, and lakes. And so we shall find that the Cumberland place-names give much evidence of the character of the different invasions. As the later invaders naturally occupied the fertile lowlands rather than the barren fastnesses of the hill tracts, we find a number of place-names of British origin in the upland regions, and they are frequent in the elevated tracts of West Cumberland. The word "combe" (Welsh *cwm*), for example, employed to indicate a half-bowl shaped hollow in the hills, is still in general use. Of British place-names we may cite "Blencathra," "Helvellyn," "Derwent," "Carlisle," "Penrith" from among a host of others.

The preponderance of Angles over Saxons among the people of the Anglo-Saxon invasion is shown by the frequent occurrence of the Anglian "ton" (originally an enclosure) and the rarity of the corresponding Saxon "ham." The latter word is chiefly found in the south of Cumberland, as at Whicham, while a number of places scattered over the lowland end in ton, as Plumpton, Skelton, Hutton.

The Danish word *by* (a village) is found abundantly in the Eden valley and the lowlands where settlements were made by the Danes who came from the east, witness Scaleby, Melmerby, Lazonby, but it also occurs in the west, as in Allonby, Flimby. Again, *thwaite* (a clearing) is one of the commonest terminations in Cumberland (Seathwaite, Rosthwaite, Stonethwaite).

Several Norse and Danish words are still in use in the local dialect for physical features, as *force* for a waterfall, *gill* for a stream, and *fell* for a hill, while yet others are in use for other things, thus a *gimmer-lamb* is a female lamb (Danish *gimmerlam*), and *smit* is used for the smear of colour with which sheep are marked.

Traces of settlements of the pre-Roman dwellers in the district have been found in many places, often on high ground. The Anglian, Saxon, Danish and Norse immigrants probably occupied some of the settlements which had been founded by their predecessors, but there is no doubt that they founded many new hamlets and villages. It has been already stated that they occupied at first the richer lowlands, and in course of time they no doubt took into cultivation the fertile floors of the upland valleys, gradually extending up these valleys into the inner recesses of the fell country.

On the arrival of the Normans the county was parcelled out into large areas divided among the Norman barons, under whom the general mass of the inhabitants lived as bondsmen; but the Norman influence was mainly restricted to the lowlands of the north and east of what is now Cumberland, and those who had penetrated into the heart of the dales remained undisturbed in the possession of their small freehold estates. Thus arose an independent set of small farmers known as estatemen, or "statesmen," having properties of from thirty to three hundred acres, which descended to the eldest child. The farm buildings were usually situated in the valleys, surrounded by fields in which oats and other crops were cultivated, or in which cattle grazed. The latter extended up the hill sides, and far above was the open fell on which the sheep grazed, and from the peat-mosses of which peat was dug for fuel. It is only within the last half-century that these statesmen have practically disappeared, and the influence of their independence in keeping up and intensifying the sturdiness due largely to the Norse blood is undeniable. The stone walls which are so marked a feature in a Cumberland dale prospect are chiefly the result of the separation of the various farm-lands and the subdivision of each farm for the different uses to which its portions were put.

The population of Cumberland, according to the census of 1901, was 266,933. Twelve English counties have a smaller population than that of Cumberland. As regards density of population, our county has on the average 177 people per square mile of land, as compared with the average of 558 per square mile for the whole of England and Wales.

It is interesting to notice that the county immediately to the south, Westmorland, has the fewest people per square mile of any English county, namely 80, while on the contrary, that to the south of Westmorland, Lancashire has with Middlesex the largest number, over 2000 per square mile. These differences of course depend chiefly upon the number and size of the industrial towns in the various counties, of which Cumberland possesses comparatively few.

14. Agriculture. Forestry.

As many of the inhabitants of Cumberland have from time immemorial essentially subsisted by agriculture, we must devote some attention to its consideration.

In doing so, it is well to note first the principal factors which control the agricultural operations of any given county or district. These are latitude, altitude, climate, soil, character of the people, and outlet for surplus produce.

Latitude is of special importance in affecting the nature of the corn crops, for the county is too far north to allow of successful cultivation of wheat on a large scale, while on the contrary it suits the growth of oats, and accordingly we find that this is the chief kind of corn grown.

Altitude is responsible for the dividing line between the areas devoted to the cultivation of crops and the permanent pasture. The area of the former is chiefly in the lowlands and valley bottoms, whereas much of the latter is on the upper portions of the fells. The line separating the tracts where arable cultivation is profitable from those of permanent pasture may in Cumberland be roughly placed at a height of 900 feet above sea-level.

It will have been gathered from the remarks made in a previous chapter that the county enjoys an equable climate with fairly warm winters and cool summers, and that it is essentially humid. Such a climate is suitable to the growth of many crops, and especially of root-crops, but has always favoured grass rather than arable farming.

We have considered the soil in the geological chapter. It was there seen that there were four important varieties, namely the clayey soil of the slate-rocks, the thin light soil of the Mountain Limestone tracts, the sandy soil of the New Red Sandstone region, and the mixture of peat and silt in the estuaries and infilled lakes. The agriculture of each of these soils is marked by special features.

The character of the inhabitants is also of great importance. The physical strength, capacity for work, and tenacity of purpose inherited from their Scandinavian ancestors and fostered by the struggle against physical conditions which cause the produce to be wrung from the earth by hard labour, have enabled the natives to overcome the difficulties with which they were confronted, when a weaker and more indolent race would have been worsted in the struggle.

Lastly, there is the outside market for the produce. In the early days corn and roots were grown in sufficient quantity for home consumption only, and the same may be said of the supply of meat. The wool of the sheep alone was carried to the markets to exchange for such necessaries of life as could not be obtained on the spot. Of recent years, owing to the expansion of large towns outside the district and the facilities for transport caused by the development of the railways, the agricultural produce of the county is taken further afield, and accordingly many cattle are reared, especially dairy shorthorns. As the result of the rearing of this stock much milk and some butter are sent to various large towns in southern Scotland and the north of England, and dairy cows and those of blue-grey breed are sent to places in all parts of the kingdom.

The following figures from the journal of the Royal Agricultural Society for 1908 give the acreage devoted to the different corn crops and the number of live stock:

Corn Crops.			*Acreage.*
Wheat	...	...	...	1088
Barley	...	...	...	1368
Oats	...	...	...	68,606

Live Stock.			*Numbers.*
Cattle	...	...	...	155,238
Sheep	...	...	...	647,410
Pigs	...	...	...	17,243

Much of the sheep pasture, as already stated, is on the higher hills, which are chiefly occupied by the old slate-rocks in south-west Cumberland, though parts of the Mountain Limestone tract with its sweet herbage also furnish excellent pasture, as do some of the higher Carboniferous rocks of the Pennine hills.

Many of the sheep are of the Herdwick breed, of which tradition says that they escaped from a vessel of the

Herdwick Sheep

Spanish Armada wrecked on the west coast of Cumberland and thence spread through the district, but they are gradually being replaced by "half-breds."

The old Cumbrian type of White Pig is getting very scarce, and the Large White (or Large Yorkshire), which has more hair and carries more lean flesh, is very largely superseding it. The most famous hams in the kingdom come from this county.

The cattle are mainly reared in the lowlands. The increased sale of milk has caused a great increase in the number of dairy shorthorns, with a diminution in the number of cattle fed for the beef-market. The latter include the famous "Blue-grey" breed, from the mating of white shorthorn bulls with Galloway cows : they come from the north and north-east of the county.

In the lowlands a certain amount of arable land exists. From the figures just given it will be seen that the growth of wheat and barley has almost entirely given place to that of oats. Of root-crops, potatoes are grown on the lighter and mangolds on the heavier soils.

Forestry is not very important in Cumberland, for less than four per cent. of the whole county is woodland. In 1300 there was one royal forest, namely Inglewood, a restricted portion of the earlier forest of Cumberland. Inglewood occupied an area of over 150 square miles, but has been long deforested. Conifers are the chief trees now grown in the county. These include larch, Scotch pine, and spruce. A considerable quantity of oak and birch is also grown.

Instruction in agriculture is provided at the Agricultural College, Aspatria, and at the Cumberland and Westmorland Farm School, Newton Rigg, near Penrith.

15. Industries.

The winning of coal and the ores of iron from the earth's interior will be considered in the following chapter. Much labour is employed in the Cleator Moor district and at Millom in the extraction of the iron from its ore, and large furnaces have been erected for the purpose.

The growth of oak, the bark of which is employed in tanning, has permitted the establishment and maintenance of a number of tanneries in various towns. They exist at Carlisle, Wigton, Penrith, Whitehaven, Workington, Soulby and Thurstonfield. The most important are at Whitehaven and Maryport.

The manufacture of textile fabrics was at one time extensively carried on, but is now practically extinct in the county.

A number of trades flourish in Carlisle, which, as the result of its commercial prosperity, has spread far beyond the old walls which once contained the entire city. The late Chancellor Ferguson mentions "iron foundries and hat manufactories, now flourishing in Carlisle," and adds that "biscuit-baking, lithographic printing, the corn and seed and bacon trades are later introductions, and have attained considerable dimensions."

Lead pencils are manufactured in Keswick. The establishment of this manufacture was due to the former supply of graphite from the Borrowdale mines.

The shipbuilding at the ports is not conducted on a large scale, but the fisheries are of some importance. The Eden and Esk fishery supplies a considerable amount of salmon as the result of netting especially in the estuary. Higher up angling is carried on rather for sport than for profit. The Derwent fishery is mainly important from the point of view of sport. The Ravenglass fishery again is of no great importance, though a certain quantity of salmon and sea-trout is taken, especially in the estuarine portion.

The Solway fisheries are of great importance, and furnish a large supply of sea-fish such as haddock, herring, mackerel, smelts, and various crustacea and shell-fish are also taken in the Solway fishery and elsewhere along the Cumbrian coast, including crabs, lobsters, prawns, oysters, mussels and cockles.

An important industry in the county, and especially in that portion which is included in the Lake District, is that concerned in providing for the tourist and visitor. Large hotels have sprung up around Derwentwater, and many smaller ones in the dales. Lodging houses also cluster thickly in some places, as at Keswick, and many people are employed in shopkeeping mainly for the visitors, and in their conveyance from place to place.

It is however to its mineral wealth and to the pursuit of agriculture that the county chiefly owes its prosperity. Of the latter we have already spoken; we may now turn to the very important industries which come under the head of mining.

16. Mining and Quarrying.

The chief mines of Cumberland are those which yield coal and iron. The coal is derived from the coal-measures of the Whitehaven coal-field, the distribution of which is shown upon the geological map. The coal-measures of this field are divided into an upper unproductive series; a middle series with seven workable seams, the principal (the Main Band) being nine feet thick; and a lower series with five inferior seams. Some of the collieries are carried a considerable distance beneath the sea. Little coal was worked here before the middle of the sixteenth century.

Wellington Coal Pit, Whitehaven

The output for 1907 (as given in the *General Report and Statistics* published by the Home Office in 1908, which contains the latest official figures) was 2,253,790 tons, having a value at the pit's mouth of £902,831, which is slightly over eight shillings per ton.

The principal iron ores occur in two areas of Carboniferous limestone separated by the Ordovician rocks of south-west Cumberland. The northern area lies around

Cleator and Egremont, south-west of the Whitehaven coal-field, and some ore has also been worked in the Skiddaw slates near here ; the southern area is situated at Millom and forms a westerly extension of the Furness district of Lancashire, though some ore is also found in an Ordovician limestone near Millom. The ore is an oxide of iron known as red haematite, occurring chiefly in fissures of the limestone.

There is evidence that the iron mines about Egremont were worked early in the twelfth century, but really important mining only started about the year 1825.

The *General Report and Statistics* above-mentioned state that Cumberland supplied 1,299,059 tons of iron ore in 1907, which ore contained 52 per cent. of available iron. The total value at the mines was £1,211,460, being somewhat under nineteen shillings per ton.

Other ores which have been extensively worked in past times are those of lead and copper, though at the present day various causes have led to the decline of the mining of both these metals in the county. The lead ore, chiefly galena or lead sulphide, occurs in two main tracts, that of the Pennine hills and that of the Ordovician rocks around Keswick. In the former tract the lead veins situated in the rocks of the Carboniferous limestone have been worked chiefly around Alston Moor, and here work has gone on since the time of the Romans. Around Keswick lead ore has been obtained in Blencathra and in the vale of Newlands and adjoining hills. Galena has also been worked extensively in the Caldbeck fells, and to some extent on the west side of Helvellyn. Most of the veins are in the Skiddaw slates, but those last named occur in the Borrowdale Volcanic series. Sulphide of copper has also been extensively mined in the Caldbeck fells and in the vale of Newlands.

Various non-metallic minerals occur in Cumberland, as gypsum which is found in the New Red Sandstone rocks of Edenside, and barytes found in several places, but the most interesting of these minerals is the graphite or plumbago of the Borrowdale "black-lead" mine near Seathwaite. This material is here associated with igneous rocks. It is nearly pure carbon, and the Borrowdale graphite has long been known for its excellent quality; but a modern process enables the inferior graphite of other regions to be utilised, and accordingly the importance of the Borrowdale graphite has declined. This graphite has long been known and used for various purposes—for medicine, dyeing, marking sheep, lining crucibles and other vessels to enable them to resist heat, and rubbing on iron to prevent rust, but the most interesting use to which it has been put is recorded by the late Chancellor Ferguson, who described a specimen which was used as a mould by a forger of coins in the reign of Henry VII. This carries far back the period at which the Borrowdale graphite was discovered.

Many good building stones are found in the county. The coarser deposits of the Skiddaw slates are locally used for houses and walls, and the Volcanic series supplies compact greenstones which are largely used in that part of the county which is occupied by these rocks. The Eskdale granite in the south is extensively quarried about

Muncaster. The rocks of the Carboniferous system, both sandstones and limestones, are locally used, and the important sandstones of the New Red Sandstone series are employed in many towns, both the lower or Penrith sandstone and the upper or St Bees sandstone contributing largely to the buildings of many towns: Carlisle and Penrith, for instance, are practically built of New Red Sandstone.

Slate of a green colour is largely worked in a seam of the Borrowdale Volcanic series which runs from Borrowdale to the Buttermere valley, the principal quarries being at Honister. Setts for paving roads are obtained from a fine-grained granitic rock at Threlkeld, the crushed stone mixed with lime is used for flags. Flags are also quarried from some of the rocks of the Skiddaw slates and Volcanic series, and road metal is furnished by many of the harder rocks.

Lime is burnt in many places where the Carboniferous Limestone is developed.

Finally, peat for fuel is dug from the turbaries or peat bogs on the higher hills in many places, and also in the flat mosses which are found on the low grounds around the estuaries.

17. History of the County.

The Roman occupation of Britain began with the visits of Julius Caesar in B.C. 55 and 54, but the Romans first arrived into what is now Cumbrian territory under

Quarrying at Honister

Agricola in A.D. 79. That event marks the beginning of the history of Cumberland, for earlier events took place in pre-historic times. For nearly four centuries the Romans dominated the district, with occasional raids by the Picts and Scots, and an insurrection of the Brigantes which was soon quelled.

The Romans first arrived from the south along the Lancashire plain, and probably after crossing the sands of Morecambe Bay, marched along the flat tracts of western Cumberland by the sea. Later they also came into Cumberland territory from the east over the hill-passes, into the Eden valley and north Cumbrian plain. They opened up the country by constructing an elaborate series of roads which will be noticed subsequently, and protected these by military camps at points of strategic importance, and under Agricola especially the Brigantes were introduced to the civilisation and luxury of the Roman conquerors.

In the year A.D. 120, as a protection against raiders from the north, Hadrian, on a visit to Britain, built the great Roman Wall which now bears his name, and runs between the Solway west of Carlisle to Wall's End on the Tyne. In addition to the military posts many small towns and settlements rose up in Cumberland, and above all these in size and importance was the Roman Luguvallium, the modern Carlisle. In the fourth century the decline of the Roman empire had begun, and at the beginning of the following century the Emperor Honorius gave to the Britons their independence.

Shortly after the departure of the Romans, who were

The Roman Wall: Birdoswald

gradually leaving the country between 410 and 430, the English invasion began by the arrival of Teutonic peoples from the country near the Elbe. There were three Teutonic tribes dwelling near the Elbe, namely the Saxons around the mouth of that river, the Angles further north, and still further north the Jutes. All these people were termed English, and as they conquered portions of our country it in turn became English. By the end of the sixth century they had annexed the lowlands of England, the Jutes occupying Kent, the Saxons a tract north and west of Kent, and the Angles the centre, east, and north-east of England.

All this time the Cumberland area remained part of an independent British kingdom extending from the Dee to the Clyde, and bounded on the east by the Pennine Hills. The kingdom was called Strathclyde and was divided into a series of smaller States, one of which—Cumbria—included what is now Cumberland, Westmorland, and North Lancashire. We know little of what went on in this region between the departure of the Romans and the arrival of the English, Danes, and Norsemen, but there were in all probability dissensions among the different States.

After the battle of Chester in 607 when Ethelfrith King of Northumbria defeated the British, southern Westmorland became English, but Cumberland and northern Westmorland remained as a British kingdom. At a later period, this tract without belonging to England was in some way subject to the English king of Northumbria, and there is no doubt that in 685, if not before that year, Cumberland had become English ground and contained many Anglo-Saxon dwellers.

In the ninth century the Danes came over the eastern passes into the northern lowlands of Cumberland, and in the following century the Norsemen arrived by sea from the west.

The details of the various invasions of the county are veiled in obscurity. Mr Burton writes: " Of these territories it can only be said, that at this period, and for long afterwards, they formed the theatre of miscellaneous confused conflicts, in which the Saxons, the Scots, and the Norsemen in turn partake. Over and over again we hear that the district is swept by the Saxon king's armies, but it did not become a part of England until after the Norman conquest."

In 945 Cumbria was conquered by the Scots and became part of the kingdom of Scotland, while what is now southern Westmorland still belonged to England.

There is much uncertainty as to the condition of Cumberland during the reign of William the Conqueror, and as to the actual date at which it was transferred from Scotland to England. In 1092, William Rufus came to the north, took possession of Cumberland and of the northern tract of Westmorland, and established the present boundary between England and Scotland, and what is now Cumberland for the first time became wholly English.

Obscure as are all the events which happened between the departure of the Romans and the arrival of the Normans their influence upon the inhabitants of the area was most marked. The people of the Brigantes had disappeared and the inhabitants had acquired those Scandinavian characters which they still possess, while the English language had become established.

We may pause here for a moment to consider an event of the utmost importance, the introduction of Christianity. Though our country was in name at least Christian during the latter part of the Roman occupation the Anglo-Saxon invasion caused a relapse to paganism—to the worship of Woden and Thor. The Danish invasion was also one of a heathen race, though the Christian religion ultimately overcame that of the Danes. While the Cumbrian area was in a state of heathendom two different Christian churches arose in the British Isles, the Celtic church to the north and west, and the Roman church to the south and east, the latter due to the influence of Pope Gregory, who sent Augustine to preach the Gospel at the end of the sixth century. The result was that the Cumbrian pagans were for some time influenced by the missionary efforts of these churches which differed on points of importance. There is little doubt that for some time the Christians in Cumberland were actually under the rule of the Celtic Abbot of the isle of Iona in Scotland. In 664 a conference of the Celtic and Romish priests was held at Whitby, and the north of England became Romish, and so remained until the time of the Reformation.

Towards the end of the reign of William II or in that of Henry I, the "land of Carlisle" was granted to Ranulf Meschyn. This land was that comprised by modern Cumberland (with the exception of Alston) and northern Westmorland. Ranulf as earl parcelled out the land into baronies, but subsequently surrendered it to the Crown. Henry then divided the land of Carlisle, the barony of Kendal, and an intervening strip into two counties, those of Carliol and Westmarieland, and entrusted their management to sheriffs, adding Alston to Carliol. In addition to the three baronies of Gilsland, Liddell, and Burgh-by-Sands made by Ranulf, Henry founded five more, namely Copeland, Allerdale, Wigton, Greystoke and Levington, reserving the city of Carlisle and the Forest of Cumberland to the Crown. In the year 1177, as stated in Chapter I, the name "county of Cumberland" was given to what was then the county of Carliol, and this name of Cumberland was ever after retained. Henry in 1133 also made the land of Carlisle into a bishopric.

After the death of Henry, Cumberland was given up to the king of Scotland, but in 1157, the third year of the reign of Henry II, it was annexed to the English Crown, and once again, and finally, became a part of England.

From this time onward the history of Cumberland is essentially that of its great border town—Carlisle. From the time of the recovery of the land of Carlisle from the Scotch by Henry II to the death of Edward I in 1307 the two nations were frequently at war, and great armies were assembled at Carlisle, but after the death of Edward at Burgh Marsh on his way to Scotland, the scene of the more active hostilities was around Berwick rather than around Carlisle, and the Carlisle region was one of raids rather than of battles. A period of active border warfare was however resumed in Cumberland in the sixteenth century, and conflicts continued at intervals until the union of the two kingdoms in 1707.

The fighting during the earlier half of the sixteenth century was between the two kingdoms, but in 1551 peace was concluded, and an attempt was made to pacify the borders, and a definite line drawn between the two countries. Part of this was artificial, marked by the construction of the Scotch Dyke.

In the seventeenth century there was much fighting during the struggle between King and Parliament.

Carlisle was again concerned in the Jacobite risings of 1715 and 1745. Little happened in 1715, but in 1745 the city surrendered to Prince Charles Edward and King James III was proclaimed at Carlisle Cross. The Highlanders marched southward but were soon compelled to retreat, and towards the end of December the Prince and his forces left Carlisle. Since then the county and its border town have been left in peace.

18. Antiquities—Prehistoric, Roman, Anglo-Saxon, Norse, Mediaeval.

Our knowledge of the history of the inhabitants of Cumberland is chiefly derived from a study of written records, but it is not entirely dependent upon it, for relics of the early inhabitants afford information even of those periods of which records were made in writing.

The tract forming the county was however inhabited in times earlier than those concerning which we have the first written records, and our knowledge of the state of the inhabitants during those early days is derived solely from an examination of relics left by them, either structures such as grave-mounds and stone-circles, or various weapons and other articles which have resisted decay and been preserved to the present day.

The periods previous to those of the first written records are usually spoken of as "Prehistoric," and we will now consider the nature of the remains which have come down to us from these times.

Neolithic Implement
(*In Kendal Museum*)

Before the use of metal for forming tools and weapons was discovered these were chiefly of stone (and in some cases of bone), and we are enabled therefore to divide prehistoric time into the Stone Age and the Prehistoric Metal Age. We will begin with the earlier of these ages—that of stone.

Antiquaries have found that there are two very different classes of stone implements marking two quite distinct ages of civilisation, of which the later was far more advanced than the earlier. We may speak therefore of the Palaeolithic or Old Stone Age, and the Neolithic or New Stone Age. In the older age the implements were formed by chipping the stone into shape, and the art of grinding and polishing them was unknown. Instruments of this type are found in the river-gravels and caverns of England as far north as Derbyshire, but are unknown in the northerly parts including Cumberland. We need

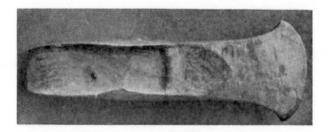

Bronze Palstave

not, therefore, dwell further upon the remains of the older Stone Age.

Many instruments of the later or Neolithic Age have been found in the county. Among them are "celts"— stones which have been chipped and ground into the form of a broad chisel with a sharp cutting edge at the broader end. These are often polished, and were probably used as hatchets, as suggested by two extremely rare finds, one made in an old Cumbrian tarn, and the other in Solway Moss, the instruments being actually found in their wooden handles. Other types are perforated hammers, hammer-axes, and trimmed flint knives. The implements are mostly fashioned from stones derived from some of the hard rocks of the district, but some, as just stated, are of flint. It is not to be supposed that all the stone implements which have been found in the county belong to the Stone Age, for stone was used long after

Long Meg and her Daughters, near Penrith

the introduction of metal, and has indeed been in use quite recently for some purposes, as for instance the "strike-a-light" used for igniting tinder.

A remarkable set of grooves was found on a rock-surface upon Lazonby Fell : these grooves are believed to have been produced by the grinding of stone implements upon the rock.

There are few structures in the county which were of a certainty made by men of the Stone Age, for there is much difficulty in distinguishing the relics of this age from that of the succeeding period. "Barrows," or burial-mounds, often largely made of stones are frequent. Of these barrows there are two types, the long and the round barrows. The long barrows belong to the early Neolithic period and the round barrows to a later period.

From study of the relics of the later Stone Age found in other places we know that the English dwellers of this age were hunters and fishermen, possessing domestic animals, and having some knowledge of agriculture. They were also acquainted with the art of making rude pottery.

The introduction of metal was certainly gradual, and long after objects formed of metal were introduced stone no doubt continued to be the principal material from which implements were fashioned. The first metal to be used for the purpose was not iron but the alloy of copper and tin which we call bronze, for the art of smelting iron was much more difficult than that of making bronze. Accordingly the earlier prehistoric age of metal was a Bronze Age. At first the Bronze Age men imitated the stone implements which were in use, and a large number of the bronze instruments are more or less modified forms of the hatchet-shaped stone instrument, but the introduction of metal allowed of the formation of a far greater variety of forms than could be fashioned in stone, and in our county, besides the hatchet-shaped forms, many bronze dagger-like weapons and swords have been discovered. Metal also lent itself to the making of ornaments, accord-

The Stone Circle near Keswick and Blencathra

ingly bronze bracelets and other ornaments have been found in the county.

The county is rich in prehistoric structures besides the barrows already mentioned. Some of these structures may belong to the Neolithic period, while others are certainly of the Bronze Age. There are several so-called "Druidical Circles," which have, however, nothing to do with the Druids. The best known are "Long Meg and her Daughters" near Little Salkeld, the circle near Keswick, and that at Swinside on the slopes of Black Combe.

Remains of settlements of prehistoric tribes are also frequent. Various camps and earthworks of pre-Roman date also occur, one of the most remarkable being that on the summit of Carrock Fell, surrounded by a low stone wall.

There is evidence that iron had been introduced into Britain before the arrival of the Romans into our island, and a few relics found in Cumberland have been assigned to this Early Iron Age.

We may now proceed to consider the relics of the Roman occupation of our land. In so doing we pass definitely from prehistoric to historic times. Of the Roman roads we shall speak elsewhere, and at the same time refer to the most important of the Roman camps.

Foremost among the Roman relics is the Roman Wall of Hadrian to which allusion has already been made. Its course from Northumberland to Carlisle is shown on the map. Comparatively few portions of the Wall in Cumberland are so well preserved as those in the less cultivated tract of Northumberland, and between Carlisle and Bowness the Wall has been practically destroyed.

It originally consisted of a stone wall, with an average width of about eight feet and a height probably of 20 feet, with a ditch on the north side. South of the Wall along a great part of its course was an earthen wall or vallum with a ditch to the south. A military road, 18 feet in width, ran between the stone wall and the earthen vallum. Large stations occur at intervals of

Stone from a Roman Ring, Castlesteads

about four miles along the Wall and, between these, small forts are found at intervals of about one Roman mile.

The chief Roman relics of small size consist of pottery of very artistic types, some of which was made in Britain, though a large part was imported from the Continent, especially from Gaul. There are also many and various ornaments, and a large number of Roman coins, which have been found in several places. Many inscribed stones

Cross at Irton

have been discovered ; some actually on the solid rock, others on quarried stone. The latter have been found in particular along the line of the Roman Wall. Many of them are tombstones.

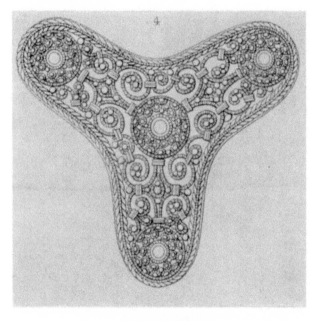

Silver Ornament Kirkoswald

There are many interesting relics belonging to the period between the departure of the Romans and the arrival of the Normans, the chief of which are sculptured stones. Some of these are of Anglian origin, while others

show Norse influence. Of the former we may notice
a cross at Bewcastle and another at Irton, and of the

The Luck ot Edenhall and its Case

latter the celebrated cross in Gosforth churchyard, whose
sculptured surface has Christian symbols along with illustrations of Norse mythology. It forms one of several

round shafted crosses, of which others occur at Beckermet, in Penrith churchyard and elsewhere.

To this period belong the curious sculptured tombstones known as "hogbacks," which are widely scattered throughout the county, and the interesting fonts of Bridekirk and Dearham churches. Of smaller objects we may notice a beautiful fibula or brooch-like ornament, found at Kirkoswald, and preserved in the British Museum. It is of silver, and ornamented with red paste jewels.

Norman relics save those associated with ecclesiastical architecture are rare, but there are several of mediaeval times ; of these we will notice only the celebrated glass vessel known as the "Luck of Edenhall" which, with its leather case, is still preserved.

19. Architecture—(a) General.

It will be convenient to consider under these headings the ecclesiastical, military, and domestic buildings of the county. Before doing so, however, we may offer some observations on the edifices in general.

We may remark at the outset that here as elsewhere the buildings are affected by the nature of the materials available, local stone being mainly used. Thus, in the area of the Ordovicean rocks, stones of that age are largely used; in the Carboniferous tracts, the limestones and sandstones belonging to that geological system have been employed ; and, in the New Red Sandstone tracts, the red sandstones which so largely constitute the beds of that period form the material for the chief edifices.

West Doorway, Calder Abbey

While treating of the materials used for buildings, it may be noted that in the localities through which the Roman Wall passes, this Wall has been used as a quarry to provide stones for all sorts of later buildings.

A preliminary word on the various styles of English architecture is necessary before we consider the churches and other important buildings of our county.

Pre-Norman or, as it is usually, though with no great certainty termed, Saxon building in England, was the work of early craftsmen with an imperfect knowledge of stone construction, who commonly used rough rubble walls, no buttresses, small semicircular or triangular arches, and square towers with what is termed "long-and-short work" at the quoins or corners. It survives almost solely in portions of small churches.

The Norman Conquest started a widespread building of massive churches and castles in the continental style called Romanesque, which in England has got the name of "Norman." They had walls of great thickness, semi-circular vaults, round-headed doors and windows, and lofty square towers.

From 1150 to 1200 the building became lighter, the arches pointed, and there was perfected the science of vaulting, by which the weight is brought upon piers and buttresses. This method of building, the "Gothic," originated from the endeavour to cover the widest and loftiest areas with the greatest economy of stone. The first English Gothic, called "Early English," from about 1180 to 1250, is characterised by slender piers (commonly of marble), lofty pointed vaults, and long, narrow, lancet-

Lanercost Priory

headed windows.   After 1250 the windows became
broader, divided up, and ornamented by patterns of
tracery, while in the vault the ribs were multiplied.   The
greatest elegance of English Gothic was reached from
1260 to 1290, at which date English sculpture was at
its highest, and art in painting, coloured glass making,
and general craftsmanship at its zenith.

Brigham Church, Cockermouth

After 1300 the structure of stone buildings began to
be overlaid with ornament, the window tracery and vault
ribs were of intricate patterns, the pinnacles and spires
loaded with crocket and ornament.   This later style is
known as "Decorated," and came to an end with the
Black Death, which stopped all building for a time.

With the changed conditions of life the type of

building changed. With curious uniformity and quickness the style called "Perpendicular"—which is unknown abroad—developed after 1360 in all parts of England and lasted with scarcely any change up to 1520. As its name implies, it is characterised by the perpendicular arrangement of the tracery and panels on walls and in windows, and it is also distinguished by the flattened arches and the square arrangement of the mouldings over them, by the elaborate vault-traceries (especially fan-vaulting), and by the use of flat roofs and towers without spires.

The mediaeval styles in England ended with the dissolution of the monasteries (1530–1540), for the Reformation checked the building of churches. There succeeded the building of manor-houses, in which the style called "Tudor" arose—distinguished by flat-headed windows, level ceilings, and panelled rooms. The ornaments of classic style were introduced under the influences of Renaissance sculpture and distinguish the "Jacobean" style, so called after James I. About this time the professional architect arose. Hitherto, building had been entirely in the hands of the builder and the craftsman.

There is one character which is common to many of the buildings of the county, whether ecclesiastical, military, or domestic, and as it is one of particular geographical import, it must be noticed. It has been seen that Cumberland as a border county has witnessed much fighting, both in actual warfare and during border raids; consequently other edifices than those erected exclusively for military purposes required fortifying. The quadrangular tower, the nucleus of the Norman fortification, served as the pattern for other fortified buildings, and accordingly the tower of this type forms a noteworthy feature of many ecclesiastical and domestic buildings.

20. Architecture — (b) Ecclesiastical: Churches and Religious Houses.

The ecclesiastical buildings of Cumberland include the cathedral of Carlisle, the churches, and the religious houses.

Carlisle Cathedral dates back to the time of William Rufus, when one Walter is said to have begun a house dedicated to the Virgin Mary, which was continued thirty years later by Athelwold, prior of St Oswald at Nostell in Yorkshire. The present west end of the cathedral contains the remains of these Norman buildings, of which the greater part was destroyed in 1646. The existing Norman part consists of two bays of the old nave and the south transept. These parts are marked by the massive columns and rounded arches. The stone used for the Norman portion was largely quarried from the old Roman Wall.

A choir on a large scale was built in the thirteenth century, being begun about 1245 and continued until near the end of the century, but soon afterwards the cathedral was burnt. Though most of the Norman work escaped, the choir save the side aisle walls and a few other portions was destroyed. The aisles are of Early English style and have pointed windows and slender, shafted

pillars. Building of the choir was resumed soon afterwards in the Decorated style of architecture, which is now
displayed in the greater part of the choir including the
beautiful east window. The north transept and the tower
have been restored in comparatively recent times, and in
1871 the fragments of the Norman nave, previously used
as a church, were thrown open to the cathedral.

Carlisle Cathedral

Apart from the stones obtained from the Roman Wall,
the building is chiefly constructed of New Red Sandstone.

The greater number of the churches of the county
are small. Many of them have undergone much restoration, and although in some cases the ancient character has
been preserved during the process, in too many it has been
completely destroyed. We may give examples of various

8—2

The Choir, Carlisle Cathedral

ecclesiastical buildings which illustrate the different styles of architecture.

In marked contrast to the neighbouring county of Northumberland, where several fine Saxon churches still exist, there is no church in our county of which the fabric is indisputably of Saxon origin, though some of the early building of Norman times must have been done by the local inhabitants, and therefore retain traces of the pre-Norman style.

As examples of Norman architecture we may mention (in addition to the portions of Carlisle Cathedral) the early part of Burgh-by-Sands church, S. of Solway Firth, the elaborate doorway of Great Salkeld, N.N.E. of Penrith, the south doorway and north side of the chancel in Torpenhow near Ireby, the nave and chancel of Dearham, E. of Maryport, and much of Over Denton, about six miles from Brampton.

Much of what remains of Lanercost Priory is of the best period of the Early English style, and the clerestory shows the change into the Early English architecture. The nave arches of Newton Reigny, the beautiful door in the north transept of Calder Abbey, and a lancet window in the chancel of Dearham church are Early English.

The architecture of the Decorated period is not well represented in the county. In addition to the portion of Carlisle Cathedral already noticed we need only mention Brigham church near Cockermouth, which according to Mr Isaac Fletcher "illustrates the phases of Gothic architecture prevalent in England from the Norman Conquest to the reign of Henry V. With the exception of Carlisle

Cathedral, it is almost the only ecclesiastical building in the county which contains a good example of Decorated architecture, and we have no other porch of that style." Greystoke church is a good example of the architecture of the Perpendicular period.

As an example of a church of the Renaissance period,

Greystoke Church

we will mention only Penrith parish church, which though of no great architectural interest is impressive.

Before passing on to enumerate the religious houses, reference must be made to a few other churches.

The church at Bolton near Wigton is in a very special sense Romanesque. It shows certain features unlike those of the other Cumbrian churches, but represented in certain churches of Scotland, and there is little doubt that the style of architecture shown was derived by the Scotch from some continental nation—probably France—with whom they were then in close association.

The churches of Burgh-by-Sands, Newton Arlosh near Holme Cultram, and Great Salkeld possess quadrangular towers, like so many other Cumbrian churches, but in the case of these three, evidence of their fortifica-

Newton Arlosh Church

tion still remains, thus illustrating the remark made in the first section of this chapter as to the community of purpose of the quadrangular towers of castle, church, and dwelling-house alike.

Turning now to the religious houses, there were Priories of the Austin Canons at Carlisle and Lanercost, of the Benedictine monks at St Bees and Wetheral, and of the nuns of that order at Armathwaite and Seton. Abbeys of the Cistercian monks were built at Holme Cultram and Calder. In addition to these there were houses of the Friars at Carlisle and Penrith, hospitals at Carlisle, Wigton, Bewcastle, Caldbeck, and St John-in-the-Vale, and colleges at Greystoke and Kirkoswald.

21. Architecture—(c) Military and other Castles.

The division into military, ecclesiastical, and domestic architecture is, for Cumberland, somewhat arbitrary. The greater number of the castles in the county were not military, though designed for defence; but this, as we have seen, was also the case with many manor-houses, and even with some of the churches.

When the Normans entered the district two important roads ran northward into Scotch territory, one by Carlisle and the other by Bewcastle, and at these points military castles were erected. They were insufficient to prevent incursions from Scotland through the intervening country, and for further protection a chain of castles was built along the line of the great road over Stainmore into the rich tract of north-eastern England. This chain consisted of castles which are situated in Westmorland, and we are here only concerned with the buildings at Carlisle and Bewcastle.

Of the latter castle there remain only an enclosure with ruins of four boundary walls, and a gateway. The exact date of its erection is doubtful. Of Carlisle Castle on the contrary much still exists, and it has played a most important part in defence at many periods ; we must therefore regard it more particularly.

In the first place let us consider the natural advantages of its site, for the Normans were quick to seize upon

Carlisle Castle, from the Interior

sites which possessed natural advantages, though, as in the present case, they frequently occupied places which the Romans with their genius for the selection of strategic positions had previously occupied.

Standing on the high ground of Carlisle one looks far away to the north across the passage over the Eden, and over the marshes towards Scotland. This high ground at one time formed two hills, on the more southern and lower of which the cathedral and town stood, while the higher and more northern was occupied by the castle. The former depression between these hills has long been filled in by gradually accumulated rubbish, and the hills form, in the strictest sense of the word, a peninsula. It is bounded on the north by the Eden, and on the east and west by tributaries, the Petteril on the east and the Caldew on the west, and as these tributaries approach very near to one another to the south of the town, a narrow constriction is thus caused, which makes the high ground almost an island. The north end of this peninsula was an admirable position for a castle.

The Norman castles as first erected consisted of quadrangular towers or keeps. Subsequently additions were made to the castles in the shape of a wall with subsidiary towers surrounding the ward or inner space on which stood the keep, while other subsidiary buildings might also be erected in this ward.

Carlisle Castle was erected by order of William Rufus, though the completion of the building was not accomplished until the days of his successors. It consisted, in accordance with the plan of erection just outlined, of a keep with subsidiary buildings. Through the long centuries it has naturally undergone much destruction as the results of warfare and decay, with partial renovations, and is now very different from what it once was, though still utilised as a garrison for soldiers. It consists at present of outer and inner wards, and within the latter is still situated the great keep.

The annexed illustration is part of an old plan of Carlisle from a drawing preserved in the British Museum, which in itself is a reduced copy of one drawn about the time of Henry VIII, and shows the general character of the castle at that date. The plan also shows part of the wall which girdled the city, with its gateways and towers. Much of the western and northern portions of this wall still stand, though altered, but the prominent round

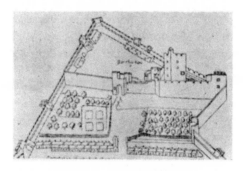

Old Plan of Carlisle Castle

buildings which attract the visitor's attention on entering the city from the railway station, though situated on the southern portion of the old wall, are of comparatively modern origin ; they occupy the site, however, of two circular towers which once existed, forming the citadel.

Between the road over Stainmore and the west coast is high ground until we reach the strip of lowland along the coast, where, as already seen, a road was carried round in the time of the Romans. Here, at a period later than the erection of Carlisle Castle, two other castles were placed to complete the system of defence. These were Cockermouth and Egremont.

Cockermouth Castle is situated on a high peninsula between two streams, where the Cocker joins the Derwent. It was begun about the end of the first quarter of the thirteenth century and additions were made at intervals for about two hundred years. It is built in general conformity with the plan of Norman castles. Egremont Castle stands on a height near the town of that name. The oldest portion was built about 1140, but there were, as usual, later additions. The great quadrangular tower is the principal portion which now remains.

Of castles not strictly military we may mention Askerton, Triermain, Naworth, Kirkoswald, Penrith, and Millom. Of the two first-named little remains. Askerton was a border fortress erected by Thomas Dacre in the reign of Henry VII or the earlier part of that of Henry VIII. It consisted of a quadrangle to hold a garrison for the defence of the barony of Gilsland. Triermain again was a border fortress built by the Vaux family. Naworth stands on the junction of two streams. It was built as a castle in 1335, but it incorporated an earlier peel tower, of a nature which will be described in the next chapter. It has been frequently altered, and still forms a residence of the Howard family. Kirkoswald Castle, standing on an eminence near the village, was founded in the twelfth century, but no Norman work is left. The principal portions were added in the fifteenth and sixteenth centuries by the Dacres.

Cockermouth Castle

Penrith Castle, of which little remains, was a rectangular building situated on an eminence west of the town. It was built in the fourteenth or fifteenth century, and differs from the other Cumbrian castles in that it was built not by a baron, but by the inhabitants themselves for their own defence. Millom Castle, in the extreme

Penrith Castle

south of the county, was begun in the thirteenth century, but the chief part now standing is sixteenth century work.

Two other buildings named castles may be mentioned here. Rose Castle, the present palace of the Bishop of Carlisle, started like Naworth as a peel tower which was incorporated with the later buildings. Dacre Castle near the foot of Ullswater, built about the beginning of the thirteenth century, is of interest inasmuch as it is of a type intermediate between the peel tower and the true castle.

Many of the castles above-mentioned were moated, and in several cases the moats or portions of them yet remain.

Dacre Castle

22. Architecture—(*d*) Domestic: Manor Houses, Cottages.

In many parts of the county, and especially in that portion which belongs to the Lake District and forms a resort for wealthy people from other places, there has been an extensive erection of modern buildings of varied styles of architecture.   With these we are not concerned, but shall consider only the more ancient buildings which present certain features typical of this part of the country.

The more ancient dwellings may be divided into two classes, the manor-houses occupied by the lords of the manors, and the cottages of the peasants.

The more interesting manor-houses were built between the fourteenth and the seventeenth centuries. The characteristic feature of these is the peel tower, rendered necessary for a people liable to border raids; it originally constituted the whole dwelling.   These towers were modelled on the keeps of the Norman castles. They were rectangular and usually three storied.   In the lowest storey were kept stores, the inhabitants occupied the middle storey by day and slept in the upper storey by night.   The roof was used for fighting purposes when raiders from the north had to be repelled.   The peel tower stood in an enclosure called the "barmkyn," surrounded by a wall.   Into this barmkyn the cattle were driven during the times of raids.   In later times dining halls and other additions were built out from the peel tower.

Such towers are found at the halls of Catterlen, Over Denton, Blencow, Hutton Hall and Hutton John near Penrith, and Isell, and in many other buildings, some of which have already been noticed.

It has been seen that after the Reformation, church building was checked, and that as far as ecclesiastical architecture is concerned, there is little to be said concerning the Cumbrian buildings of the Renaissance period.

Catterlen Hall

It is, however, otherwise with the manor-houses, which continued to be set up during this period. Many of them were erected in an L-shape, and the peel tower disappeared. The classic influence is displayed in many of the details of these buildings. As examples of houses of the Renaissance period we may notice two halls near Cockermouth. The first of these, Huthwaite Hall, was ostensibly built in 1581, though parts may be older. The windows have a decided Elizabethan character. The second house, Ribton Hall, was built in the time of Charles II.

The earlier cottages of the peasants were usually built of rough stone, often without mortar. They generally had three rooms on the ground floor, a sitting room and kitchen combined, a parlour, and a dairy. Stone steps, which were often outside the house, led to a loft or sleeping-room. Various modifications naturally occur.

Under the head of domestic architecture we may mention the bridges, many of which are characterised by extreme simplicity of construction. This is specially so with bridges carrying old pack roads and byways over streams.

23. Communications—Past and Present.

Important roads connect the Cumbrian lowlands with the south, east, and to some extent the north. From the south and east, any road other than one going round the Cumbrian coast must cross the uplands which form the two arms of the ⊣ described in Chapter 7; to do so they take advantage of the various passes.

The Brigantes no doubt possessed, like latter-day barbarians of other countries, an intricate network of paths connecting hamlet with hamlet. Such paths would be kept open through the undergrowth of the lowlands,

Stockley Bridge, Borrowdale

and would in many cases extend over the higher ridges. Having no definite construction, those subsequently abandoned would tend to disappear, while those which continued in use would show no signs of their formation during pre-Roman days.

There is some evidence of two such pre-Roman routes which were afterwards utilised. One came from

9—2

York to Carlisle, and the other from Lancaster over the Pennines into Northumberland. Further details as to their direction will be given when we trace the Roman roads.

The object of the main Roman roads of the county was to connect the northern plain of Cumberland with the south and also with the east. At first Chester was the important military centre from which expeditions set out for Cumberland, but with the rise of York the decay of Chester took place, and subsequently the road entering the Cumbrian tract from the south-east became of prime importance.

We will consider first the southern approach. As stated in Chapter 17, the first expedition (that of Agricola in A.D. 79) was probably around the west coast. Passing over the sands from near Lancaster to the Furness coast, and thence over Duddon sands this road would enter the county near Millom, and from thence continue along the low ground of the coast to Bowness-on-Solway. Along this line are Roman camps at Ravenglass, Moresby (2½ miles N.E. of Whitehaven), Ellenborough near Maryport, and Mowbray. At Bowness the road continued past Carlisle along the line of country which was after-wards occupied by the Roman Wall. Camps or stations existed among other places at Bowness, Drumburgh, and Burgh on the way to Carlisle (Luguvallium), and east of Carlisle at Castlesteads, and at Birdoswald (Amboglanna), where is the second largest fort on the Wall. East of Birdoswald the Wall and the route of the old road pass into Northumberland.

At a later period, when the Romans had occupied

much of the territory, shorter routes to Cumberland would be followed. Starting from Lancaster, one went by way of Windermere and over the Dunmail Raise Pass into Cumberland, and then past Keswick, and on to the Roman station near Wigton. Another from the Lancashire area went up the Lune and over Shap Fells, and down the Eden valley to the camp near Kirkby Thore in Westmorland. Just north of this it entered Cumberland, and was carried over the Pennines to Alston, beyond which it passed into what is now Northumbrian territory, turned west, re-entered Cumberland near Birdoswald, and on reaching that station turned northward past Bewcastle into Scotland.

Another road came from York (Eboracum) over the pass of Stainmore into Westmorland, and north of Brougham (Brovacum) entered Cumberland, passing through the New Red Sandstone lowlands to Carlisle. The last Roman road which we need notice communicated with the west coast of Cumberland. It left the road from Lancaster to Wigton at the head of Windermere where there is a camp. Passing up the Brathay, and over Wrynose Pass and entering Cumbrian ground, it crossed the Duddon and went over Hardknott Pass where a camp occurs perched high on the hillside, and down Eskdale to the important station and Roman fort at Ravenglass.

After the wane of Roman influence, there were no important changes in the means of communication until railways were introduced. The main Roman routes were still trodden, with occasional slight deviations, and the important passes of the Pennine-Cheviot hills and through

the Lake District watershed still dominated the main
lines of traffic.

A network of new routes was formed as the area
became more populous. They probably began as foot-
paths and bridle-tracks, and some by degrees were used
by vehicles.

Even in the case of some of the roads used by wheeled
traffic, the difference from the old Roman roads is usually
easily detected, for the former are often badly graded,
and it is only here and there that they depart from
the old tracks. There are however many exceptions,
especially the so-called military road between Newcastle
and Carlisle, constructed in the middle of the eighteenth
century, and the well-graded road from Hexham to
Penrith by way of Alston, constructed by the celebrated
road-maker MacAdam, at the beginning of the second
quarter of last century.

The old pack-roads were used for carrying produce
from the farms and hamlets to the nearest market towns.
Such as have not been converted into modern roads have
in many cases disappeared. As the valley bottoms were
often thickly covered with undergrowth which could with
difficulty be penetrated, the pack-roads were frequently
taken up the sides of the fells from the houses in the
valley-floors, and accordingly we still meet with relics of
these pack-roads on the fells where there is now little or
no traffic; they usually took advantage of the smaller
passes which indent the ridges between valley and
valley.

The year 1838 witnessed the completion of the first

railway in Cumberland, namely that from Newcastle to Carlisle. Since that time Carlisle has become a most important railway-centre, and the railways can best be treated by taking first those which radiate from Carlisle like seven spokes of a wheel.

The Newcastle and Carlisle line, which now belongs to the North Eastern Company, comes through a pass between the Tyne and Irthing, enters the county, and runs through the Irthing valley into the Vale of Eden.

Taking the other important railways in order, and going round a circle in the direction of movement of the hands of a clock, the next great line is the Midland, which comes down the Vale of Eden from the Pennine Passes near its source, entering the county near Newbiggin.

The London and North Western line is, in Cumberland, parallel to the Midland, and a few miles to the west of it. It enters the county near Penrith and goes by way of the Petteril valley to Carlisle.

The Maryport and Carlisle line runs between those towns over the low ground of New Red Sandstone.

The North British Railway sends a branch from Carlisle to Silloth and Port Carlisle, north of the last-named line.

The Caledonian leaves Carlisle for the north, and enters into Scotch soil at Gretna junction.

The North British has another branch from Carlisle past Longtown. This enters Scotland near Riddings.

The most important railway which does not reach Carlisle is that of the Furness Company. Leaving the North Western at Carnforth, it makes a rough semicircle round the south-western side of the Lake District. It enters the county south-west of Broughton-in-Furness, and terminates at Whitehaven.

The Cockermouth, Keswick and Penrith railway runs between Penrith and Cockermouth, going over the high pass of Troutbeck between Penrith and Keswick.

The London and North Western Company has some important lines linking those of other companies in the north-west of the county. One from Whitehaven to Maryport connects the Furness and the Maryport and Carlisle lines, and from this is a branch from Workington to Cockermouth.

The Caledonian has a branch connecting Scotland with west Cumberland, having acquired the Solway Junction Railway, which crosses the Solway by a viaduct near Bowness, a mile and a quarter in length.

A branch of the North Eastern line from Haltwistle in Northumberland to Alston enters Cumberland close to the latter town.

In that portion of the county which is largely frequented by tourists, coaches are still much used for their conveyance, but motor vehicles are now used for public conveyance of passengers on some routes, as for instance on the Keswick-Windermere road, and they will no doubt become much more frequent.

In 1819–23 a canal was made from Port Carlisle to the city of Carlisle, but it never paid.

A steamer plies on Ullswater during the season.

It is interesting to notice that notwithstanding changes in the method of travel, the principal routes are in the main those of the old Roman roads, which were skilfully planned so as to take the lines of least resistance as regards the barriers formed by physical features.

24. Administration and Divisions— Ancient and Modern.

It was noted in an earlier chapter that Cumberland was definitely formed into a county by Henry I, but that before the Norman invasion the Saxons had divided tracts in southern England into shires. These shires were "shorn" off from larger tracts of country for administrative purposes.

In the early days of tribes administration was largely a family affair and later an affair of clans, but as the tracts of land under one ruler increased it was necessary to have some definite divisions, in each of which administration of some of its affairs was local, and the administration according to families and clans was replaced by a territorial one. The early Saxon "shires" of the south, each probably the result of expansion outwards of a definite colony, formed convenient territorial areas for administrative purposes, and under the Normans, those areas which had not thus been parcelled out were formed into counties. Hence the existence of the county of Cumberland for similar purposes.

The Saxons had *ealdremen* or governors who appointed deputies called *sheriffs* (shire-reeves; reeve being equivalent to our word steward). The inferior people were partly *ceorls* or freemen, and partly *villeins* who were labourers in the service of particular persons, and not strictly slaves. Upon the establishment of the feudal system by the Normans many of the Saxon laws and customs were retained, as was also the old distinction of classes. Thus there were counts or earls, barons, knights, esquires, free tenants, and villeins. When Henry I formed the separate counties of Cumberland and Westmorland the administrative power was put into the hands of Sheriffs, for Ranulf Meschyn ruled over the land of Carlisle as an earl, but it was found inadvisable to leave the control of the borders to a man with the powers possessed by the earls.

As seen in Chapter 17, the county was divided in the time of Henry I into eight baronies. The baronies and the Forest of Inglewood (a restricted portion of the Forest of Carlisle) were further subdivided into manors.

Those who were responsible for the administration of the affairs of the county saw to the collection of taxes for the Crown, the supply of soldiers for military service, and the administration of justice within the county. It will thus be seen that the county is an area which was separated for the purposes of internal government, but it also took some part in the affairs of the nation.

The system of local government gave rise to still smaller divisions for administrative purposes. The county was early divided into five *wards* (a term which is peculiar to the border counties). The wards were those of Allerdale-above-Derwent, Allerdale-below-Derwent, Cumberland ward, Leath ward and Eskdale ward. There are now seven, the two additional being termed the Derwent and Bootle wards. A further division was made into *parishes*, each with its own officials, and the parishes were again subdivided into *townships* or *constable-wicks*. As the shire had its sheriff, so the parish had its own special reeve, or presiding official.

The gradual accumulation of numbers of people in restricted areas giving rise to towns necessitated special government in the case of these towns apart from the constable of an ordinary constablewick with his subordinates.

The city of Carlisle must have had some sort of municipal government in the twelfth century, and in a document of the year 1292 a mayor of Carlisle is mentioned, though it is doubtful whether he was recognised by the Crown ; but a charter of 1353 shows the existence of a recognised mayor and bailiffs, the city being then perfectly independent of the county and of all county jurisdiction, and being in all but name a county in itself. The city is now governed under a charter of Charles I, which incorporates the governing body under the name of the "mayor, aldermen, bailiffs, and citizens of the city of Carlisle." This body was to consist of twelve aldermen, one of whom was to be appointed mayor, two bailiffs, two coroners, and twenty-four capital citizens. In 1835 this body was dissolved and a new one established consisting of a mayor and nine other aldermen and thirty councillors.

There are two other corporate towns in Cumberland, namely Whitehaven and Workington. Whitehaven was incorporated in 1894. The Town Council consists of a mayor, six aldermen, and eighteen councillors. Workington received its charter of incorporation in 1887, and has a mayor, seven aldermen, and twenty-one councillors.

When the county was first constituted its voice in the general affairs of the nation was slight. When the great Charter was signed in 1215 one of its provisions was that its articles were to be carried out by twelve sworn knights from each shire chosen in the County Court. Thus the influence of the county in national affairs became more direct. In 1295 the first complete Parliament assembled, and, besides others, two knights were summoned from each shire, two citizens from each city, and two burgesses from each borough. Since then Cumberland has had its full share in the government of the nation, and after many changes it is now represented by six members of Parliament chosen by the burgesses of the following divisions and boroughs, each of which returns one member:—Northern or Eskdale Division, Mid or Penrith Division, Cockermouth Division, Western or Egremont Division, Borough of Carlisle, Borough of Whitehaven.

Let us turn now to the present government of the county, which has gradually grown out of the old administrative system. The head officer of the county is the Lord Lieutenant, who in some ways represents his Norman predecessor who under the title of count, earl, or other name, was at the head of affairs. The Lord Lieutenant represents the Crown in the county, and one of his duties is to nominate all Deputy Lieutenants and Justices of the Peace.

The High Sheriff of to-day is also to some extent representative of the Norman Sheriff, although his duties are much restricted in comparison with those of the ancient officer, and are largely connected with affairs of the law. He is Keeper of the King's Peace within the county, and he attends the judges of the realm when on circuit.

A recently constituted body with purely administrative functions is the County Council. It consists of a Chairman, Vice-Chairman, Aldermen and elected Councillors. The county is divided into nearly seventy electoral districts, each of which returns a councillor : about one-third of this number are made aldermen. Among its functions are the management of county halls and buildings, pauper lunatic asylums, bridges and main roads, the appointment of certain officers such as coroners, the control of parliamentary polling districts and contagious diseases of animals. The County Council are also the local education authorities through an Education Committee. There are also Rural District Councils and Urban Councils, for the administration of smaller areas of the county.

For purposes of Justice the county has Assizes which are held at Carlisle, Quarter Sessions presided over by a Chairman of Quarter Sessions, and also a number of Petty Sessions, each having Justices of the Peace, whose duty it is to try and to punish offenders against the law.

For Ecclesiastical purposes, Cumberland (except Alston) with Westmorland and part of Lancashire forms the diocese of Carlisle, which is divided into three archdeaconries and further subdivided into nineteen rural deaneries, eleven of which are in Cumberland. A still further division is made into ecclesiastical parishes,

which are not the same as the civil parishes.   There are
167 of the former and 213 of the latter parishes in the
county.

25. The Roll of Honour of the County.

Foremost among those who have done honour to their
county must be placed the great families of whom individuals have through the centuries been prominent for
promoting the welfare and directing the affairs of the
county, and in many cases rendering important services
to the country.   The Aglionbys, Curwens, Grahams,
Howards, Lowthers, Musgraves, and Penningtons have
rendered such services.   It would be clearly impossible
in a small work on geography to enumerate the numerous
services of different individuals of these families, and we
must be content to make this general reference.

The selection of notable men is difficult: no two
writers would agree as to the list of worthies to be included in the roll of honour.   The present list has been
chosen, partly with the intention of illustrating the contributions of prominent Cumbrians to administrative work,
warfare, law, religion, literature, science, art, and industry.
These names are included not always so much on account
of their eminence as from regard to what they have done
for the county.   In two cases names have been included
of men who were not born in the county, for to the
geographer, the accident of birth is not of so great
importance as the selection of the county for residence
on account of particular advantages which it may offer.

We may place side by side the names of two men who had much in common, John Christian Curwen and Sir James R. G. Graham. Distinguished as parliamentary representatives, they did much for the promotion of agriculture in the county, which therefore owes much to them.

William Wordsworth

John Christian (1756–1828), who married a Curwen and afterwards took the name of Curwen, was born at Ewanrigg in the parish of Dearham. He was devoted to the progress of agriculture and has been spoken of as the "father of Cumberland agriculture." Sir J. R. G. Graham of Netherby (1792–1861) is described in the *Victoria County History* as "perhaps the most illustrious parliamentary figure the county of Cumberland has ever produced."

Among lawyers one name stands out prominently—that of Lord Ellenborough. Edward Law (1750–1818),

Wordsworth's House, Cockermouth

who became first Baron Ellenborough, was born in the parish of Dearham, and became Lord Chief Justice of England.

In a county possessing a cathedral city, the list of ecclesiastics of varying degrees of eminence is naturally large, but as their connection with the county is in a geographical sense accidental, it is unnecessary to notice

individuals.   The three names selected for consideration
are those of men of eminence who have been born in the
county.   All were raised to the episcopal bench, but for
very different acquirements.

Robert Southey

E. Grindal (1519?–1583) was son of a farmer of
Hensingham near St Bees.   He became Archbishop of
Canterbury, and is noted rather for his position than for
any great ability other than was concerned in his advance-
ment, for he made little mark.   William Nicholson, born

in 1655, was son of the rector of Plumbland. He became Bishop of Carlisle and was subsequently made Archbishop of Cashel, but died before taking possession of the see. He is known especially as a diarist, and as one who was interested in history, archaeology, and botany. George Mandell Creighton (1843–1901) was born in Carlisle.

Greta Hall: the home of Southey

He was chiefly noteworthy as a historian, and became Bishop of Peterborough and subsequently of London.

Of men who have made their mark in warfare it would be easy to give a long list referring to the early days of border strife. This list would include names of those great families of whom we have spoken. In more peaceful modern times no very great soldier or sailor can be claimed as Cumbrian.

Among those who have advanced science we may note Thomas Addison (1793–1860). By accident of birth he was Northumbrian though of Cumberland family. He was a physician, eminent as a teacher, who also did much to raise the reputation of Guy's Hospital in London. He diagnosed a disease which is called after him. The great scientific light of Cumberland was John Dalton (1766–1844), the distinguished chemist, noted for his "Atomic Theory," which has gained him a high place

John Dalton

among scientific worthies throughout the world. He was born at Eaglesfield near Cockermouth the son of a Quaker weaver. In contrast to Dalton who shone as a theoretical chemist was H. L. Pattinson (1796–1858) who applied his knowledge of chemistry to the manufactures. He is best known as the originator of "Pattinson's Process" for extracting silver from lead-ores. The particular direction in which he applied his powers was no doubt due to his birthplace, Alston, the centre of a lead-mining district.

Archaeology lies on the borderland of science and literature. Many archaeologists have naturally arisen in a county so rich in relics of our predecessors. Prominent among these not only for the extent of his knowledge and the variety of his attainments but for the solid contributions which he has made to the history of his county from a study of antiquities and documents alike, was Richard S. Ferguson (1837–1900), who was born in Carlisle and became Chancellor of the Diocese.

As contributors to art we may name L. Watson (1804–1847), sculptor, who was born about six miles from Carlisle in the Caldew valley, and Samuel Bough (1822–1878), born at Carlisle, an artist of very high rank, many of whose pictures portray local subjects.

Lastly we have to consider the contributors to literature, and here our roll of honour is brightly illumined, for Cumberland claims with its sister county the honour of possessing the members of the Lakes School of Poets, who clustered around the central figure, Wordsworth.

William Wordsworth (1770–1850) was born at Cockermouth. It is true that he lived 'the greater part of his life in Westmorland, but his writings are as strongly influenced by the county of his birth as by that of his later residence. Great as a poet, he must also be regarded as one who has contributed largely to the prosperity of the county, for the recognition of the Lake District as a place of beauty owes much to his writings.

Around Wordsworth gathered others: Samuel Taylor Coleridge, Hartley Coleridge, Thomas De Quincey, and Robert Southey. Of these Robert Southey (1774–1843) may be claimed as a Cumbrian though not by birth, in that he chose Keswick as his place of residence during the latter part of his life, and that some of his writings are influenced by the character of his surroundings in the county.

The Birthplace of John Dalton, Eaglesfield

26. THE CHIEF TOWNS AND VILLAGES OF CUMBERLAND.

(The figures in brackets give the population in 1901. M.B.= Municipal Borough, U.D.=Urban District. Those not lettered are Civil Parishes. The figures at the end of each section are references to the pages in the text.)

Alston (3134). A market town on the eastern side of the Pennine hills, 29 miles east-south-east of Carlisle. It is situated among moors, on a steep hill near the junction of the rivers Nent and South Tyne, at a height of 960 feet above sea-level. The church is dedicated to St Augustine: the existing building is modern. The town possesses a town hall, a grammar school, and a high school. The mining villages of Nenthead and Garrigill are in the neighbourhood. Alston is the centre of the lead-mining district of Alston Moor. The mines belonged to the Earl of Derwentwater and after his execution were made over to Greenwich Hospital. In connection with these mines is an aqueduct known as Nent Force,—a subterranean canal five miles in length which was cut by the Trustees of Greenwich Hospital. Natural caverns are found in the limestone, one of which is known as Tutman's Hole. The Roman Maiden Way passed through here. (pp. 11, 38, 133, 136, 147.)

Arlecdon and **Frizington**, U.D. (5341, Arlecdon 1632, Frizington 3709). The town of Frizington and the neighbouring village of Arlecdon are situated to the east of Whitehaven. The inhabitants are concerned with iron mining and to some extent with agriculture. An iron furnace was erected at Frizington about the middle of the eighteenth century.

Arthuret (2455). In this parish are situated the villages of Longtown and Netherby. They lie to the north of Carlisle. Longtown, on the high road to Edinburgh, was once a market town, but has lost much of its former importance. Arthuret church (St Michael's) was built in 1609. It contains many memorials of the Graham family. Netherby Hall or Castle, the seat of the Grahams, is on the east side of the river Esk. It is built around a peel tower. (p. 135.)

Aspatria with **Brayton**, U.D. (2885). Aspatria is named from Gospatrick, father of the first lord of Allerdale. It is situated between Wigton and Maryport. St Kentigern's church has been rebuilt in recent years, and the castle has disappeared. There are coal-mines in the parish. Hayton Castle, a former seat of the Musgraves, is now a farm. The Agricultural College was founded in 1874. (p. 83.)

Beckermet St Bridget (555) and **Beckermet St John** (516) lie south of Egremont, not far from the west coast, in an agricultural district. Calder Bridge village and Calder Abbey are in the parish. The old church of St Bridget dating from the thirteenth century is now used as a mortuary chapel. In the churchyard are two crosses of pre-Norman times.

Bewcastle (700), lies between two feeders of the river Lyne, close to the northern boundary of the county. It is of interest on account of its antiquities. A Roman camp has yielded many altars and other relics. The castle (of doubtful date) is almost entirely destroyed, an enclosure with remains of four walls and a gateway only being left. The restored church of St Cuthbert is near the Roman camp. It possesses the celebrated pre-Norman Cross. (pp. 107, 120, 132.)

Bolton (898), lies to the south of Wigton. The inhabitants are mainly devoted to agriculture. All Saints' Church at Bolton Gate is an ancient edifice of the Romanesque style. (p. 118.)

Calder Abbey

Bootle (759), near the southern end of the county, is situated on a gentle slope between Black Combe and the sea. It is an old market town. The church of St Michael's, mostly rebuilt, contains a brass of Sir H. Askew (1562). The ruins of Seton nunnery one mile north of Bootle have lancet-shaped windows. The view from Black Combe is very extensive. On the summit was a beacon which existed so far back as the fifteenth century. (p. 36.)

Bowness (1079) is on the Solway, west-north-west of Carlisle. Here ended the Roman Wall, and a camp was formed in connection with it. The church, which is largely rebuilt, contains an ancient font. Port Carlisle is in the parish. Drumburgh castle, an old manor-house, is now a farm. At Bowness was once a ford across the Solway to Scotland, and the viaduct of the Solway Junction Railway crosses the estuary here. (pp. 103, 132, 136.)

Brampton (2494), nine and a half miles east-north-east of Carlisle, lies between the rivers Irthing and Gelt, south of the Roman Wall. It was a Roman station. The soil around Brampton is fertile. The town itself is ancient. There is a large market place with a town hall (built in 1817) in the centre. Brampton is situated in a district peculiarly rich in antiquities. The old church of St Martin, one and a quarter mile from the town, is built partly of stones from the Roman Wall, and has a lancet window. Naworth Castle, two miles to the east of the town, is the old baronial seat of the Lords of Gilsland.   It was burnt down in 1844 and rebuilt, so that but little of the ancient edifice exists. Close by is the beautiful ruin of Lanercost Priory. In the neighbourhood also are Askerton and Triermain Castles, and Over Denton Church with Norman architecture. From Brampton too may be visited Gilsland with its Spa, and the camp of Birdoswald (Amboglanna) the second largest camp on the Wall, from which many relics have been obtained.

Brigham (723), west of Cockermouth is a village whose inhabitants are occupied with agriculture, coal-mining, and quarrying. The importance of its church has already been noticed. (pp. 117, 119.)

Broughton (1334) is situated in the Whitehaven coal-field between Cockermouth and Workington. The inhabitants are chiefly engaged in coal-mining.

Burgh-by-Sands (844), about five miles north-west of Carlisle, and south of the estuary of the Eden, is in a rich agricultural district. It is situated upon the Roman Wall, and a camp stood here. On the site of this camp is the fortified church to which allusion has been made. A monument near the village marks the spot where Edward I died in 1307. (pp. 117, 132.)

Caldbeck (863) a village on the north side of the Caldbeck Fells, where mining has been extensively carried on; and where many rare minerals occur. The church of St Mungo at Caldbeck has some Norman work, but is chiefly restored. Hesket New-market on the Caldew lies to the south; it is a town which has lost its former importance.

Carlisle, M.B. (43,480) is an episcopal city and a municipal and parliamentary borough. It is doubtful whether the Britons had any settlement here before the Romans came, but in Roman times the town of Luguvallium occupied part of the site of the present city, and a Roman stockade of three rows of oak posts has been discovered. For about two centuries after the departure of the Romans Carlisle became a British town, but towards the end of the seventh century Ecgfrith made Carlisle English ground, though, as we have seen, it was not yet part of the English kingdom. In the eighth century the town was burnt by the Danes. During the two centuries which elapsed between the English occupation and the arrival of the Normans Carlisle became "British or nothing." William II has been called the founder of Carlisle, but as Mr Freeman says "he was a founder only on ground where others had been founders long before him," hence Carlisle bears a British name—*Caer Luel*.

We have already touched upon some of the more important historical events of later times in the chapter on the history of the county.

Notwithstanding its history, or rather because of it, Carlisle possesses few old buildings. The castle, the old walls and the cathedral have already been noticed. The castle is a dépôt of the Border regiment. Some remains of the priory still exist partly incorporated in the deanery. The existing churches, so far as their materials are concerned, are of modern date.

The town hall is not an imposing building; it was erected in 1717. The so-called guild-hall in the Green Market is a half-timbered structure of the fourteenth century. The market-cross in the market-place was erected in 1682. Tullie House, a seventeenth century building, contains a public library and museum and in the latter is a valuable collection of local objects. In the Tullie House is also the Bibliotheca Jacksoniana, a collection of books relating to the district.

The grammar school was founded by Henry VIII. The present buildings in Cuthbert's Close were finished in 1883.

The two large circular towers near the railway station are the Assize Courts, and the prison adjoins them. They have been standing for about a century. The fine sandstone bridge over the Eden was completed in 1815.

The part of Carlisle devoted to manufactures is chiefly outside the walls, especially on the western side. (pp. 3—5, 24, 32, 38, 42, 76, 84, 91, 96, 97, 103, 114, 119—23, 132—6, 134—41, 146, 148.)

Cleator Moor, U.D. (8120): a modern town north of Egremont. The adjacent village of Cleator is old, with a restored church.   Cleator Moor is a town of importance owing to the haematite mines and iron furnaces.   There is also a flax mill here.   (pp. 45, 83, 87.)

Bridekirk Font

Cockermouth, U.D. (5355), at the junction of the rivers Derwent and Cocker, is a market-town and former parliamentary borough.   On a hill called Papcastle on the right bank of the

Derwent about a mile below the junction of the Cocker is a Roman station which has been largely utilised as a quarry for building Cockermouth. The castle has already been noticed. All Saints' church is a modern structure. Bridekirk is about two miles distant. (pp. 23, 25, 44, 124, 130, 136, 137, 147, 148.)

Crosscanonby (931) is close to the west coast, and a little north of Maryport. The church of St John has some Norman work, an old font, and some pre-Norman sculptures.

Dacre (886). The village of Dacre is about five miles west-south-west of Penrith, not far from the lower end of Ullswater. Dacre Castle, the old seat of the Dacres, has been noticed in an earlier chapter. The ancient church of St Andrew has been largely restored: in it is a red sandstone effigy of a man in chain armour said to date from the thirteenth century. In the parish of Dacre is the village of Great Blencowe with a school founded by Thomas Burbank in the reign of Queen Elizabeth, and a hall with a fifteenth century peel tower. (pp. 125, 126.)

Dalston (1925) lies four miles south-south-west of Carlisle and is situated in a fertile tract of pasture and arable land. The hall has been enlarged but contains a peel tower. The grammar school foundation probably dates from the time of Elizabeth.

Dearham (2147) east of Maryport is a colliery village. Its church, with much Norman work and a south doorway to the nave of Transitional Norman and an Early English lancet window in the chancel, has been restored. It contains an ancient font. (pp. 108, 117, 143.)

Distington (1922) about four and a half miles north-north-west of Whitehaven is a colliery village. Near it are the ruins of a manor house known as Hayes Castle.

Egremont, U.D. (5761, Egremont 3599, Moor Row and Bigrigg 2162). Egremont is a market-town on the river Ehen situated in the iron-ore district of west Cumberland. The church of St Mary is of ancient foundation, but the fabric is modern. Little remains of the castle of which the oldest portion dates back to about the year 1140. (pp. 23, 87, 124.)

Farlam (1365) is situated near the north end of the Pennines not far from Brampton. In addition to agriculture and quarrying, the inhabitants of the parish are occupied in coal-mining, for coal occurs in the neighbourhood. Talkin Tarn is near.

Flimby (2482) is a village two and a half miles south-east of Maryport which has grown extensively during the last half century, for the inhabitants are largely employed in the coal-mines.

Gosforth (935) is a west Cumberland village on the road from Seascale to Wastwater, whose inhabitants are chiefly engaged in agricultural pursuits. In the church and churchyard are many pre-Norman remains including the celebrated Gosforth Cross. Gosforth Hall is a seventeenth century building, with a fine ingle-nook. (p. 107.)

Great Clifton (1029) is on the Derwent two and a half miles east of Workington. The parish is in the Whitehaven coal-field and coal is mined in it.

Harrington, U.D. (3679) is a modern town with harbour which owes its importance to the coal-trade. There are several coal-pits in the parish, and also iron works. The church has a font of the twelfth century. Near by is a Roman camp. (pp. 34, 36.)

Hayton (1216) is two and a half miles south-west of Brampton. The inhabitants are occupied with agriculture and quarrying.

Hensingham (2090) is a village one mile south-east of Whitehaven, of which town it may be regarded as a suburb.

Hesket-in-the-Forest (1860) owes its name to its situation in the old forest of Inglewood. The parish lies between Penrith and Carlisle, and contains several villages.

Gosforth Cross

Holme Cultram, U.D. (4275 of whom 2393 are in the civil parish of Holme Low). This district lies in a projecting mass of land between Moricambe Bay and the sea. In addition to several inland villages it contains the watering places of Silloth and Skinburness. St Mary's Abbey, Holme Cultram, is of red sandstone and was founded in the twelfth century. Only a portion of the nave is left. The west door is a late example of round-headed

Keswick and Derwentwater from Latrigg

Gothic work, the nave arcade is more advanced. It contains the figure of an abbot with mitred head seated on a throne. (p. 135.)

Keswick, U.D. (4451), an old market-town on the river Greta situated at the foot of Skiddaw, about half a mile from Derwentwater. It is the "metropolis" of the Lake District and the only town in that part of Cumberland which is situated in the district. It was once of importance as a mining centre, and as the result of the occurrence of graphite in Borrowdale still has lead-pencil works, but it now depends upon the influx of tourists and is a town of hotels and lodging houses.   St John's Church on the outskirts of the town was built in 1838 but the church of St Kentigern is at Crosthwaite some little distance from the town. It has some Perpendicular work, and contains male and female effigies probably of the fifteenth century.   The stone circle is on an eminence to the east of the town.   (pp. 84, 85, 87, 103, 133, 136, 149.)

Lamplugh (1119) is a scattered village between Cockermouth and Egremont.   Limestone quarries occur in the neighbourhood.

Maryport, U.D. (11,897) is a modern town owing its importance to the coal and iron fields of west Cumberland. Coal has been worked here since 1750 and iron furnaces were erected about the middle of the eighteenth century.   The harbour has grown apace and possesses large floating docks.   In addition to the coal and iron trade the town has shipbuilding works, tanneries and various other industries.

The Roman camp at Ellenborough is situated upon a hill overlooking the sea.   A very large number of Roman remains—altars, inscribed stones and other relics—have been at various times discovered.   (pp. 33, 42, 84, 132, 135, 136.)

Millom, U.D. (9182) is a modern town situated upon the west side of the estuary of the Duddon at the southern end of the county.   It owes its importance to the rich deposits of haematite iron ore contained in the limestone.   The church dedicated to the Holy Trinity is old, but has suffered much from restoration. It contains effigies of a gentleman and lady of the Hudleston family on an elaborate tomb of the fifteenth century.   Millom Castle, the seat of the Hudlestons, dates from the thirteenth century but was rebuilt in the following century, and the tower is a sixteenth century erection.   It is situated on low ground to the north of the town.   Objects of interest in the neighbourhood of Millom are the stone circle at Swinside on the eastern slopes of Black Combe, the old bloomery or iron furnace of Duddon Bridge, and a red sandstone effigy of a lady in Whitbeck Church.　(pp. 5, 11, 12, 23, 29, 33, 36, 84, 87, 124, 126, 132.)

Roman Stele: Plumpton Wall

Moresby (1056), a colliery village two miles north of Whitehaven. There is a Roman camp here, and many Roman relics have been found.　(p. 132.)

Parton (1406) a village on the sea coast north of White-haven. Its day as a port is now over and the inhabitants are engaged in fishing, or employed in engineering works and a brewery.

Penrith, U.D. (9182) is an old market-town and the centre of a rich agricultural district. It is built in a valley between the red sandstone Beacon Hill and the limestone Red Hills, not far

Penrith Parish Church

from the Eamont. The castle is the relic of a rectangular building of red sandstone doubtfully referred to the fourteenth century. It stands on a height west of the town. The house of the Austin Friars has disappeared. Gerard Lowther's House (Two Lions Inn) in Dockwray Street has a fine ceiling in the parlour with coats of arms, put up in 1585, the house itself being older.

The church of St Andrew is a Renaissance red sandstone

building erected 1720–22. In the churchyard are two pre-Norman crosses and four hogbacks re-arranged to form the "Giant's Grave." The neighbourhood is rich in objects of antiquarian and archaeological interest including the Roman camp of Plumpton Wall, the peel tower of Catterlen Hall, a seat of the Vaux family, in addition to many nearer objects on the Westmorland side of the Eamont. The Beacon Hill has a tower erected in 1719 : a beacon certainly existed here in the fifteenth century. (pp. 5, 7, 23, 29, 32, 38, 42, 76, 83, 84, 108, 120, 124, 126, 128, 134—6.)

Ravenglass, a small town at the junction of the rivers Esk, Mite, and Irt on the triple estuary. It is interesting for its scenery and antiquities. A Roman camp and harbour were here, and a road led to the camp on Hardknott. A Roman villa " Walls Castle," 52 feet × 43 feet built of red sandstone, is near. The harbour is now silted up and the town decayed. A beacon existed on Muncaster Fell in the fifteenth century. The gullery noted in the chapter on Natural History is near Ravenglass. (pp. 36, 45, 84, 132, 133.)

St Bees (1236), a village near the west coast, close to the sandstone Head of the same name. A nunnery was established here by an Irish saint, Bega, about the middle of the seventh century, but it was destroyed by the Danes. William Meschyn founded a priory in the twelfth century, which in the following century was burnt by the Scots, but the church was spared and still exists. The grammar school was founded by Archbishop Grindal in 1587. The bathing place is on the bay, some little distance from the village. A beacon formerly stood on the Head. (pp. 12, 34, 36, 89, 119, 145.)

Seascale (356) is a growing watering place north of Ravenglass, with golf links and coaches to Ennerdale and Wastwater in summer. It is more important than the census figures suggest, owing to the number of summer visitors. (p. 35.)

Seaton (1594) is near the coast, north of Workington, and on the opposite side of the Derwent. Fire bricks are manufactured here.

Stanwix (4052) is a parish with many villages. Of these Stanwix is a suburb of Carlisle on the north side of the Eden. Here was a station upon the Roman Wall.

Wetheral (3293), a large parish containing several villages. That of Wetheral is on the banks of the Eden about 4½ miles

Walls Castle, Ravenglass

east-south-east of Carlisle. Wetheral priory was founded in 1088 by Ranulf Meschyn. The tower only remains. Corby Castle near Wetheral is on the banks of the Eden and is celebrated for its beautiful grounds. The house has been rebuilt around a peel tower. (p. 119.)

Whitehaven, M.B. (19,324) is a seaport, market-town, municipal and parliamentary borough at the southern end of the

Cumbrian coal-field, on the coast north of St Bees Head. In the reign of Queen Elizabeth it was only a small fishing hamlet with six houses. It owes its prosperity chiefly to the coal trade, but also to some extent to the iron mines and works. The coal was first worked here about 1620 and Whitehaven was beginning to export coal in 1660 and then owned a fleet of forty-six vessels. The coal industry was developed by the Lowther family. In 1778 the American privateer, Paul Jones, attempted the capture of the town but failed.

The town is situated around a small inlet, in a hollow of the hills which here back the coast line. It is mostly modern, and built with much regularity, most of the streets being at right angles. The fine harbour is protected by two long piles. (pp. 33, 84, 85, 136, 139, 140.)

Wigton, U.D. (3692) a small market-town eleven miles south-south-west of Carlisle. It is in the centre of an agricultural district. Near by is the Roman camp of Old Carlisle, the materials of which were largely used for building the old houses of Wigton. The church was built towards the end of the eighteenth century in the Classical style. (pp. 84, 120, 133.)

Workington, M.B. (26,143) owes its importance to the iron trade. It is situated at the mouth of the Derwent. The town is essentially modern. At Borough Walls on the north are the remains of a Roman station. Workington Hall, the seat of the Curwens, has been almost entirely rebuilt in the last 200 years, but traces of an old peel tower remain. In the church of St Michael's, which possesses little of the old fabric, are effigies of Sir Christopher Curwen and his wife Elizabeth de Hudleston dating from 1450. On Workington Hill was formerly a beacon.

Coal was worked here in 1650, but the trade has greatly diminished of recent years, and the manufacture of iron is the staple industry. The harbour is by no means large for the requirements of the port. (pp. 34, 42, 84, 136, 139.)

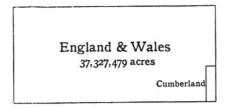

Fig. 1. The Area of Cumberland, excluding water (961,544 acres), compared with that of England and Wales

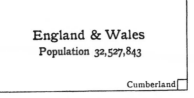

Fig. 2. The Population of Cumberland (266,933) compared with that of England and Wales in 1901

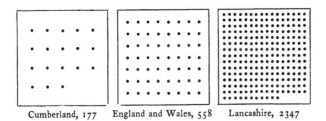

Cumberland, 177   England and Wales, 558   Lancashire, 2347

Fig. 3. Comparative Density of the Population to the square mile in 1901

(Each dot represents 10 persons)

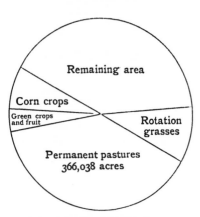

Fig. 4. Proportion of Permanent Pasture to other Areas in Cumberland (1908)

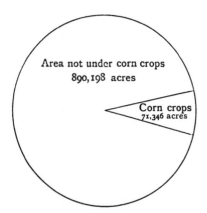

Fig. 5. Proportionate Area under Corn Crops in Cumberland (1908)

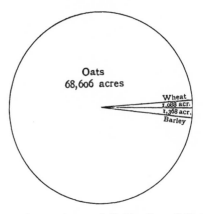

Fig. 6. Proportionate Areas of Cultivation of Oats, Wheat, and Barley in Cumberland (1908)

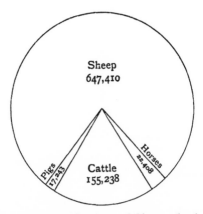

Fig. 7. Proportionate Numbers of Sheep, Cattle, Horses, and Pigs in Cumberland (1908)

9 781107 660199